Social Sanctuaries 思 姚仁喜 | 大元建筑作品 30×30
KRIS YAO | ARTECH SELECTED WORKS

30×30

KRIS YAO ARTECH | 姚仁喜 大元 建筑工场

姚仁喜 著

辽宁科学技术出版社

This three-volume monograph contains 30 selected projects spanning 30 years of KRIS YAO | ARTECH's works. This monograph is a testimony of our efforts and devotion in constructing contemporary architecture – shaping sense of place, developing space drama, and elevating cultural context. The monograph is organized in the following three sections:

Cultural Scenes exhibits the projects that emphasize celebration of cultural and historical roots,
Communal Forums focuses on the spaces that inspire creativity in our pulsating society, and,
Social Sanctuaries presents works that conveys inherent tranquility for reflection and contemplation.

Summer, 2015

本套书精选姚仁喜 | 大元建筑工场成立 30 年来多种类型的 30 件作品，分别以三大系列综合呈现，记录我们对于构筑当代形式、经营场所精神、发挥空间戏剧与提升文化意涵的努力。

"艺" 空间经堂入奥，以坚实的构筑实体，彰显历史文化的人文氛围；
"聚" 空间着重于人之聚合，关照并提供常民生活的舞台；
"思" 空间借由静谧建筑的力量，构筑安定、沉净的心灵场所。

2015 夏

Kris Yao has demonstrated his humanity with the development of physical elements
that hold true to human emotion and conditions,
cultural and historical context, and a sense of scale and place.
His architecture has a poetic nature,
using his native eastern aesthetic and spirituality
with a sense of natural light,
interplay of surfaces and forms and executing all with a high level of innovation and professionalism.

Commendation by the American Institute of Architects (AIA)
at the 2014 Honorary Fellowship Awards Ceremony

姚仁喜建筑师利用建筑元素的创作,
具体掌握了人类情感与生命状态、
文化语境与历史涵构、场所精神与人性尺度,
充分彰显了他的人文精神。
他以来自东方的美学与心灵的涵养,
以素材、造型与自然光的交互辉映,
加上高度的创意与专业的执行力,
建筑因而盈溢诗意。

美国建筑师协会 2014 年于芝加哥颁发姚仁喜建筑师荣誉院士之颂词

CONTENTS 目录

QUICKENING AND SLOWING: THE GENIUS OF KRIS YAO'S SINGULAR MODERNISM

Michael Speaks, Ph.D.

Dean and Professor, Syracuse University School of Architecture

Our bus exited the freeway onto a residential street that led, after a short distance, to the entrance of the Water-Moon Monastery (2012). Standing in front of an immense reflecting pool and framed by the roof, column, and wall lines of his own composition, architect Kris Yao, dressed entirely in black, greeted us and welcomed us onto the grounds of the recently completed Zen Buddhist retreat. A dozen or so of us had been traveling by bus for more than a week visiting buildings in Shanghai and in Taiwan, with the ultimate purpose of awarding the 8th Far Eastern Architectural Design Awards. This was among the last stops on our itinerary and I was grateful that the frenetic architecture pilgrimage was drawing to a close in this tranquil and masterfully composed complex of buildings laid out on the Guandu plain between the Keelung River and the Datun Mountains.

Upon entering the monastery grounds, we passed through a series of sound-squelching concrete walls before proceeding into the capacious Main Hall where sunlight, air, and even the muffled freeway noise is filtered, tempered and made part of Yao's composition. Though a monastery, Water-Moon is not a sanctuary from the world but is rather a sieve through which the external world of the city — its smells, sounds, tastes and touches — passes before entering the internal world of the mind and of contemplation. The hall, in fact, is both a central circulation junction where visitors cross paths with worshipers and nuns, and a wondrous sensory mixing chamber where the controlled stream of noise and chaos from the city is reconditioned and made to blend with the sound of prayers, visitor's chatter, and the soft shuffle of those moving about on concrete floors. A magnificent teak box hovers above the main hall, blocking the sun's rays and framing views into and out of the open, transparent expanse. The western edge of the Main Hall's second level is defined by a long, thickened wooden wall that projects sunlight through apertures carved in the form of Chinese characters. These sun-projected "light characters," which are part of the 260-word "Heart Sutra," move imperceptibly along the surface of simple wooden and concrete walls and around unadorned concrete columns, seeming to slow the very passage of time itself.

Outside, and to the east of the Main Hall, a large courtyard is punctuated by a series of irregularly situated black boulders that recall another era, if not another age. One is left to wonder whether the architect placed these massive rocks in the courtyard, or whether the courtyard was designed around them to accommodate glacial, rather than human, placement. Are they ornamental architectural details or ancestral traces of a more fundamental, geological plane on which the architect laid out his plane of composition?[1]Pre-cast concrete panel screens perforated by apertures carved as Chinese characters identify the nun's second story quarters and sit atop a concrete column corridor that frames the courtyard before wrapping around to define the Monastery's eastern edge. One can only imagine a similar movement of Chinese characters across the interior walls of the nun's quarters to that which occurs in the Main Hall. Only here the nuns are witness to, and intimate participants in, this slowing of time and recanting — with light — of the entire 5000-word "Diamond Sutra" each morning as they rise. When one experiences, as they surely do, this cinematic projection across the walls and columns of their living quarters, the registration of a slower, thickened time becomes ritual and defines a different rhythm of life.

Water-Moon Monastery is an uncompromisingly modern temple designed for a progressive and enlightened client: the Dharma Drum Mountain Buddhist Foundation. Following principles set out by DDMBF founder, Zen Master Sheng Yen, Yao designed a temple that is spare and stark, but one that is visually rich and luxurious to the touch; that is free of excess, ornament and color, but that inspires one to contemplate all that exceeds the human self. Yao transforms the Guandu plain into a plane of composition on which he skillfully arranges vertical and horizontal planar surfaces made of concrete, wood, earth and water, so as to filter and transform our experience of the contemporary world — a chaos that appears and disappears at such speed that we humans literally cannot fully apprehend it. Water-Moon Monastery slows this pace of appearance and disappearance in order to create a state of contemplation that aspires to an absolute speed that only a pantheistic god of the kind Spinoza imagined, or an enlightened follower of Master Sheng Yen, could attain. Consistent with Zen Buddhist practice, Yao thus slows our normative perception to enable a non-normative perception that aspires to survey and comprehend at absolute speed, the very speed at which the entirety of the world itself appears and disappears.

Yao slows our perceptions but he also simultaneously registers, in his composition, the co-existence of many different temporalities — from his use of large black rocks in the plaza to index geological time, to the use of water-jet cut concrete panels to index the ways in which he uses contemporary digital fabrication to translate ancient scriptures into a text of light using only sunlight, wood and concrete surfaces. The many digital photographs sent via email to Yao by nuns and visitors alike, picturing reflections of the scriptures and of the Buddha on the rain-soaked stone pavement, in the reflecting pool, and in the sky, evidence the varying speeds of modernization all occurring simultaneously on the multiple planes of Yao's remarkable composition. Water-Moon is a working monastery, but it has also become a site much visited by tourists and others seeking tranquility and temporary respite from the noise and chaos of the city. Indeed, what is perhaps most impressive is that Yao has so skillfully managed to co-mingle, yet keep distinct, these very different worlds — the world of the city and the world of the temple — to the great mutual benefit of both. And in so doing he has created an entirely new temple typology, one that is neither cluttered with the commercial and personal bric-a-brac found and often left behind in other temples, nor contrived, precious or artisanal, as might easily have been the case in the hands of a lesser architect.

Water-Moon Monastery is a masterpiece, but it is also emblematic of Yao's distinctive approach to Modernism. If Modernism is one form of architectural response to modernization — post-modernism, critical regionalism, high-tech and parametricism, are others — then Yao's work not only exemplifies a Modernism different from that which has arrived in Taiwan and Asia from the West, but it also demands a more complex definition and understanding of modernization itself. Rem Koolhaas, Yao's collaborator on the Taipei Performing Arts Center (2015), has suggested in his 2014 Venice Biennale catalogue that modernization began to accelerate in 1914 with the onset of the first world war, transforming national identity, and thus national architecture identity, into a universal Modernism that today, more than 100 years later, can be seen in cities around the world. Cities, Koolhaas suggests, and the buildings that define them, which in 1914 looked very different from one another, today look very much alike. Koolhaas's assumption is that modernization begins in the West and quickly colonizes the rest of the world, leaving only "non-architectural" building practices and customs as traces of national architectural identity. Modernization thus creates, as its byproduct, a universal Modernism, a stylistic paste made from the pulverized remnants of national architecture identity, which is spread evenly around the world by large corporate architecture firms, pushed even into the most underdeveloped economic crevices, cracks and deformations, to create a uniformly smooth surface. Evidence of this spread has given adherents and antagonists alike, reason to believe that globalization is the completion of a linear process that began in the West and that has smoothed over and re-made the rest of the world in its image and likeness.

What globalization has instead revealed is that modernization is not homogeneous and it is not linear. Modernization does not begin in the West and spread around the world, transforming difference into sameness. Rather, global modernization is heterogeneous and non-linear and is defined neither by space nor by time, but instead by speed. Any survey of any global city will reveal a fractal urban fabric of pockets, bands and swirls, all modernizing unevenly. Kris Yao works in the seams of these pockets, bands and swirls, slowing and quickening the pace of modernization in order to respond to the constraints and opportunities presented by the project. There is perhaps no better example of this than the Lanyang Museum (2010) in Yilan, on the northeastern coast of Taiwan. In a formerly active port area in Black Stone (Wushih) harbor, which has been reclaimed as wetlands, Yao designed what is now one of the most commercially successful museums in Taiwan. Lanyang Museum is designed in the sharp, angular vocabulary of local cuesta rocks that were formed over millennia by oceanic erosion. In fact, it is hard to distinguish the museum from the cluster of cuesta rocks out of which it emerges, situated today, as they have been for millennia. Articulated as alternating layers of glass for public areas and cast aluminum and granite panels for museum gallery spaces, Lanyang Museum is not so much a sieve that filters out the world as it is an enormous framing device through which visitors view the surrounding mountains, plains and ocean, whether outwardly through the glass frames or inwardly as framed exhibition galleries. Here, as in the Water-Moon Monastery, Yao creates a place of respite and contemplation. But the ambitions of the museum, which are to educate and entertain, are not the same as those of the temple, which is to enlighten. At Lanyang Museum, which was founded to showcase the natural beauty and cultural riches of Yilan County, Yao introduces us to a marbleized temporality that appears to us all at once: looking out from the glass frame of the museum across the harbor and towards Turtle Island, the distinction of "before" and "after" the founding of the commercial

harbor in the Qing Dynasty, "before" and "after" human settlement of this part of Taiwan, disappear in the mist and we are left to ponder a world where all of time and all of culture become part of the same temporality. Here, natural and man-made formations, geological and archaeological time, and aboriginal, Chinese, Dutch, Spanish and Japanese culture are larded together in order to provide the fullest and most robust experience of Yilan County.

In his 2002 Exhibition at the 8th International Venice Architecture Biennale, Kris Yao staged a simulacrum of his then uncompleted High Speed Rail Hsinchu Station, hung with photographs of two passengers who view one another across the platform, making momentary, though meaningful, connection, before boarding trains traveling in opposite directions. In this project and in other recent museum and cultural buildings, Yao is concerned not only with the materiality of architecture, with how they filter or frame the outside world, but also with our emotional experience of these spaces and places, with the ways in which we encounter and relate to each other as we transition through the multiple and complex temporalities of global modernization: from the corporate boardroom to the archaeological site; from the museum to the high speed train station; and from the freeway to the monastery, and back again to the freeway. Having spent several hours in the summer of 2014 at the Water-Moon Monastery, I, along with the other jurors from the 8th Far Eastern Architectural Design Awards, departed in a more relaxed state than when we arrived. As the bus drove off towards the freeway, I caught a final glimpse of the monastery reflected, on the pond, in near perfect symmetry. And though I could not claim to have seen the temple reflection with the perfect clarity that the Zen masters are said to be able to achieve, I had seen enough to make up my mind about which project would receive the only vote I would cast for first prize.

[1] See Quentin Meillassoux, After Finitude: An Essay on the Necessity of Contingency (London, 2010). In Chapter 1, Meillassoux describes pre-human reality as "ancestral."

序 文

快速 · 缓行：姚仁喜独特的
现代主义之精髓

麦可 · 史毕克斯（Michael Speaks）博士

美国雪城大学建筑学院教授及院长

我们搭乘的巴士下了快速道路，进入住宅区的街道上，短距离后便抵达农禅寺水月道场（2012 年）的入口。站立在一面巨大的倒影水池前，在屋顶、柱列、墙面的线条所框出的空间，一身着黑的姚仁喜建筑师在他自己创造的场景中招呼我们，欢迎我们来到甫完工的佛教禅修静地。我们十余人过去一周在上海、台湾四处参访建筑物，为了遴选第 8 届远东建筑奖的得主，终于来到行程的最后几站了。这趟建筑"朝圣"之旅的行程非常紧张，来到位于基隆河与大屯山中间关渡平原上的这座静谧而精心构筑的建筑中，我非常开心行程即将接近尾声。

进入禅寺庭园，我们经过一系列隔音的清水混凝土墙之后，踏进了宽敞的大殿。在此，阳光、空气甚至快速道路的噪音都被过滤、梳理，洗练成为姚仁喜作品的一部分。虽然实为禅寺，水月道场并非是远离世界的庇护所，而是一个过滤器，将城市外在世界的声、香、味、触经过筛滤，才进入我们心灵与静观的内心世界。事实上，大殿既是动线交汇处，访客在此与信众或比丘尼擦肩而过；也像是一个融合各种感知的奇妙音箱，将筛滤过的城市喧嚣与浑沌重新组合，并与诵经声、访客的细语声、在水泥地上走动的柔软摩擦声全部融为一体。宏伟的柚木盒子空悬于大殿之上，遮蔽了直射的阳光，却又在底部提供透明无垠的视野框景。大殿二楼西侧是一堵厚实的木制长墙，上面刻着 260 字的"心经"，光线透过镂刻的文字洒进来。这些阳光所投射的"光字"，不着痕迹地沿着极简的木墙移动，亦或环绕着朴实的圆形石柱旋转，时间似乎因而静静趋缓。

大殿东侧外面庭院，有一些不规则的黑色巨石点缀其中，令人忆起一个时代或另一段时空。是建筑师将这些庞然巨石安置于此，还是庭院设计围绕着这些非人造、状似远古巨石而安排的？这些石头是建筑师作为装点的细节，还是古老、原始的祖先遗迹，建筑师只是在其上结合构成的？[1]镂空了经文的预铸清水混凝土长墙由清水廊柱撑起，墙后是比丘尼的寮房，界定了大殿的东侧。我们可以想象：一如大殿，这些经文光字在比丘尼寮房的墙上移动的光景。每天黎明拂晓，只有住在这里的比丘尼，才能以朗诵整部 5000 多"光字"的"金刚经"，亲身见证到纾缓的时间。当此电影影像般的光影投射在寮房墙面与梁柱上时，比丘尼们或任何人，对于这种缓慢而浓缩的时间所转化成的仪式性，以及因而定义出不同的日常生活步调，一定会有极为深切的体会。

水月道场是不折不扣的现代寺宇设计，业主为思想开明先进的法鼓山佛教基金会。秉承了创办人圣严法师所阐述的理念，姚仁喜的设计简单朴素，却仍充满视觉的美感与丰富的质感，同时又低调大气，没有过度的装饰与色彩，是一座令人凝思超越人类自我格局的建筑。姚仁喜将关渡平原摇身一变，巧妙安排清水混凝土、柚木、水面与大地，精炼地构成垂直面与水平面，进而过滤并转换了人类难以领会的即刻生灭之浑沌经验。水月道场将这种生灭的速度放慢，创造出一种禅思的情境，让我们置身其中，向往只有斯宾诺莎想象的泛神，或圣严法师的开悟弟子才可能达到的绝对速度。姚仁喜的作品符合了禅宗的信念，减缓了我们的凡俗感知，让我们一探非凡俗的领域，驱使我们探究并理解绝对速度，那也正是整体世界生灭的速度。

姚仁喜虽然减缓我们的感知，但他也同时让许多不同的"时间性"共存于建筑中。从广场黑色大石头遥指着远古地质年代，一直到清水混凝土墙以水刀切割，仅借助阳光、柚木与混凝土表面，将古经文翻译成光文字指出的现代数位技术。完工以后，姚仁喜收到许多寺院比丘尼或访客传来的照片，其中许多是经文或佛像映照在雨后的石板地上、反射在水池中或倒映在天空中的景象；这些景象都见证了现代化不同的速度，都呈现在姚仁喜非凡的作品的多元层次之中。水月道场既是个禅寺，但同时也是许多访客的景点，以及寻求远离城市喧嚣与混乱、渴望短暂安宁者的庇护所。事实上，这也许正是姚仁喜最令人赞叹之处，他巧妙地将两个迥然不同的世界既区分又融合，一是世俗的城市世界，一是肃穆的禅寺世界，使其互惠共存，相辅相成。以此，他创造出全新的寺庙型态，既不像那些塞满商业和个人小古玩的一般佛寺，也不会像纯熟度不足的建筑师一般，掉入矫揉做作、过于精致或艺品化的陷阱。

水月道场不仅是一件杰作，也是姚仁喜实践他独特的现代主义之象征。如果说现代主义是建筑对于现代化的一种回应，而后现代主义、批判性地域主义、高科技与参数化主义等是对现代化的其他回应的话，那么姚仁喜的作品不仅体现了不同于从西方传到中国和亚洲的现代主义，更让我们了解，对现代化本身应该有更

深刻的理解与定义。与姚仁喜联手合作台北表演艺术中心（2015年）的雷姆·库哈斯，在2014年威尼斯双年展中提出，现代化于1914年随着第一次世界大战的开始而加速，不仅改变了民族认同，进而也改变了民族建筑的认同；因此，在100多年后的今日，世界各地的城市都可见到普世的现代主义。库哈斯认为，城市与定义城市的建筑物，在1914年看起来差异相当大，如今却都看似相同。库哈斯的假设是：现代化始于西方，并且迅速地移植到全球各地，只留下一些"非建筑"的建设与习俗，算是民族建筑认同仅存的痕迹而已。因此，现代化创造出一个普世现代主义的副产品，一种从民族建筑认同残渣所制造出来的风格拼贴，透过大型的企业型建筑师，将之推布于世界上各个角落，甚至连最未开发的经济体之缝隙、裂痕与形变都不放过，营造出均匀平整的表面。这种扩散使得支持或抵抗全球化的人都不得不相信，全球化是始于西方的线性完整过程，而且已经把整个世界的其他地区处理平滑，重塑成与其相似的意象了。

然而，全球化所揭示的却是：现代化并非同质不变，也非线性。现代化并非始于西方而后传播至世界各地，将差异转化成千篇一律。相反地，全球现代化是异质性、非线性的；既非由空间也非由时间界定，而是由速度定义。在任何全球城市中调查，都会显示城市肌理中的各种角落、条带或漩涡中，各都呈现不均匀的现代化进程。姚仁喜就在这种角落、条带或漩涡的接缝中，或减缓、或加速现代化的步调，以对应每个建筑案所带来的限制与机会。在此，或许没有比兰阳博物馆（2010年）更好的例子了。这栋在台湾宜兰东北海岸的建筑，坐落于以前的乌石港，现在则是再生的湿地。姚仁喜在此设计了台湾当今票房最成功的博物馆之一。兰阳博物馆以单面山岩石锐利棱角的词汇设计，那是经过千年海洋侵蚀而形成的独特造型。事实上，博物馆与那些环绕四周、已经存在数千年的单面山石很难区分。兰阳博物馆以玻璃营造的公共区域，以及铸铝板与花岗岩形塑的展示空间相互交错；它并非在筛滤世界，而是一幅巨大的框景，经由此，访客可观赏环绕四周的群山、平原与海洋，无论是透过玻璃框架由内向外望，还是由外向内观赏框景内的展示厅，都是如此。与水月道场相同，姚仁喜在此创造了一个庇护、沉思的地方。然而博物馆的目的是为了寓教于乐，而禅寺是为了度化众生，理念并不相同，兰阳博物馆始建是为了展示宜兰的自然美景与丰富文化，因而，姚仁喜向

我们全然地展现了凝冻时空的淬炼：从博物馆的大玻璃窗向外望，视线朝向港口及远处的龟山岛，清朝作为商业港口"之前"与"之后"的分别，以及人类居住于此地的"之前"与"之后"的分别，这些差异似乎都在云雾中消逝，令人不禁思考所有时间、所有文化皆成为同一"时间性"的一部分。在此，大自然与人造构造、地质与考古时间、原住民、汉族、荷兰、西班牙、日本文化在此都迭合为一，提供了最丰润、最道地的宜兰体验。

姚仁喜在 2002 年第 8 届国际威尼斯建筑双年展中，以当时尚未完成的高铁新竹站为题，制作了一个类月台的装置展示。一张照片上有两名乘客面对面地站在月台的两边，在各自上车乘往相反方向之前，产生瞬间却别具意义的人与人之连结。在此作品以及近期如兰阳博物馆等文化建筑中，姚仁喜不仅关注建筑的物质性，如何过滤或框围外在世界的同时，他更关注我们在这些空间场所的情感经验，关注在穿越全球现代化多重复杂的"时间性"时，人与人之间的相互联系：无论是在企业董事会议室或考古遗址、博物馆或高铁站、高速公路或禅宗寺庙，都是如此。在 2014 年夏日的那一天，我与随行的其他建筑奖评审在水月道场驻留了数个小时；当我们离开时，心情已经比抵达时更为放松。当巴士驶向快速道路，我回望了水月道场最后一眼，它倒映在池水中，近乎完美对称。虽然我不能声称自己一如禅师，能以明心慧眼见到完美无瑕的禅寺倒映，但我已经清晰地在脑海中，看见我将投给首奖的唯一一票。

¹ 参见甘丹·梅亚苏，《有限之后：论偶然的必然性》（2010 年伦敦）。第 1 章中，梅亚苏将前人类现实称之为"祖先的"。

Social 思 Sanctuaries

HEART OF TRANQUILITY

Kris Yao

The Great Learning Chapter in The Book of Rites states: "Understanding stillness, then you are able to abide; abiding leads to tranquility; tranquility leads to peace, peace leads to deep contemplation, and contemplation leads to attainment."

Over the years, I've come to realize that a building can subtly transform people's state of mind. A tranquil building can calm our cluttered, chaotic minds, and help us settle and contemplate. However, such a building cannot be designed by an agitated mind, which is precisely why Eastern cultures emphasize self-cultivation as the core of the arts.

In the recent years, I have adopted a more intuitive design approach. This approach has undoubtedly reduced the disturbances of fluttering thoughts. The power of tranquility has allowed me to be observant of the noises and insecurities in my mind, and has allowed my pervasive contrivance and doubts to gradually settle.

静谧的心

姚仁喜

"知止而后有定，定而后能静，静而后能安，安而后能虑，虑而后能得。"《礼记·大学》

建筑能潜移默化人们心的状态，这是我深信的。一座静谧的建筑，能安定我们杂乱的心，让我们深沉、思考。然而，静谧的建筑无法以喧闹的心境设计出来，因此，东方文化强调修持一己之心，才是从事艺术的核心。

近年来，我多以"直观"做设计，这个方法似乎可以减少繁杂思绪的干扰。此外，安静的力量则让我更能观察心念的喧哗或不安，使我们心中无所不在的造作与怀疑得以逐渐地沉淀。

LUMINARY BUDDHIST CENTER

养慧学苑

Taichung, Taiwan, China | Completion 1998

中国 台湾 台中 | 1998 年完工

2015 摄

2015 摄

LOCATION	Taichung, Taiwan, China
CLIENT	Luminary Buddhist Society
FLOOR LEVELS	8 Floors, 2 Basements
BUILDING STRUCTURE	Reinforced Concrete
MATERIALS	Exposed Aggregate Finish, Glass Unit Masonry, Black Cast-iron Baluster
BUILDING USE	Religious
SITE AREA	418 ㎡
TOTAL FLOOR AREA	2627 ㎡
DESIGN INITIATIVE	1995
COMPLETION	1998

项目位址	中国 台湾 台中
业主	香光尼僧团
楼层	地上 8 层、地下 2 层
建筑结构	钢筋混凝土
材料	洗石子、玻璃砖、黑色铸铁栏杆
用途	宗教
基地面积	418 ㎡
总楼地板面积	2627 ㎡
设计起始时间	1995
完工时间	1998

1998 摄

Main entry, viewing the Buddha from the street
主入口，于街道上可望见位于三楼的佛像

2015 摄

The Luminary Buddhist is a modern urban Buddhist temple, with spaces for Buddhist ceremonies, teachings, gatherings, and residences for nuns. Since its completion in 1998, the center has become an archetype for new spiritual centers in Taiwan and draws many interested visitors with its unique architecture.

The center is located on a typical storefront lot (13.5 meters wide and 34 meters deep) with buildings surrounding its three sides. It is designed with an introspective focus to create a tranquil place for meditation. The project transforms the traditional horizontal courtyard sequence of a temple into a vertical one through a central atrium that provides natural light to the spaces within. Visitors explore the courtyard on multiple paths, just as one would explore a traditional temple. The shrine, unlike the traditional configuration, is located on the third floor. It can be seen directly from the street via the main entrance and courtyard.

Drawing inspiration from Buddhist philosophy, the design seeks to retreat from the surrounding urban chaos through a space conducive to introspection. The heavy, monolithic pebble stone facade blocks out urban distractions while the randomly placed glass-block openings imply the presence of a central courtyard within.

养慧学苑是一所现代都会佛寺，空间可供礼佛、讲学、聚会和尼姑住所等用途。自 1998 年完工以来，此学苑已成为台湾新寺院的原型，许多访客也因独特的建筑设计前来感受宁静而庄严的精神场域。

养慧学苑属于典型的店面街屋型态 (13.5 米宽，34 米深)，三面围绕着其他建筑物。设计以内观的概念，建造适合禅修的宁静空间，并且把传统寺庙的水平进深中庭序列改造为垂直向度的中庭空间秩序，为室内引进自然光。访客可从不同的路线观赏中庭景致，如同参观传统寺庙一样。大殿不同于传统配置，位于学苑的三楼，从街道上透过主要出入口和中庭即能直接望见。

受到佛教思想的启发，本案设计目的在于构筑一处远离纷攘尘嚣的内观空间。外观采用厚重的洗石子立面阻绝都市纷沓扰攘，错落的开窗手法，透过玻璃镶嵌洞孔暗示涵纳的中庭 。

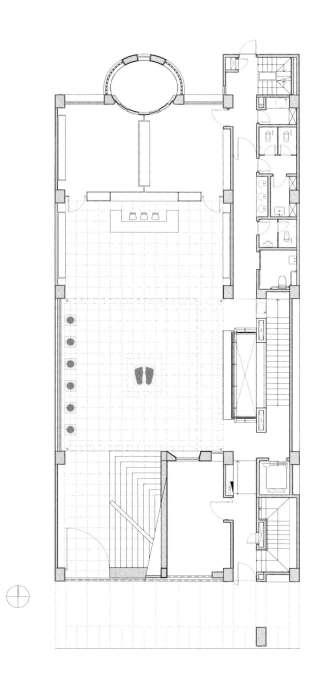

Ground floor plan
地面层平面图

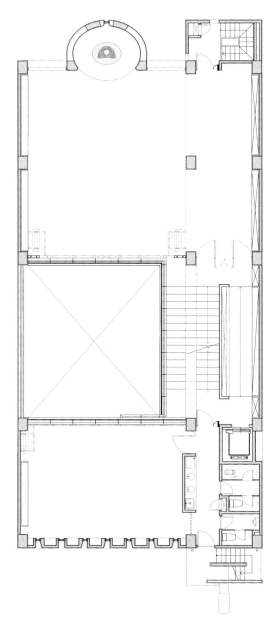

Third floor plan
三层平面图

1998 摄

Shrine with wooden doors and windows
佛堂及中庭周边实木门窗

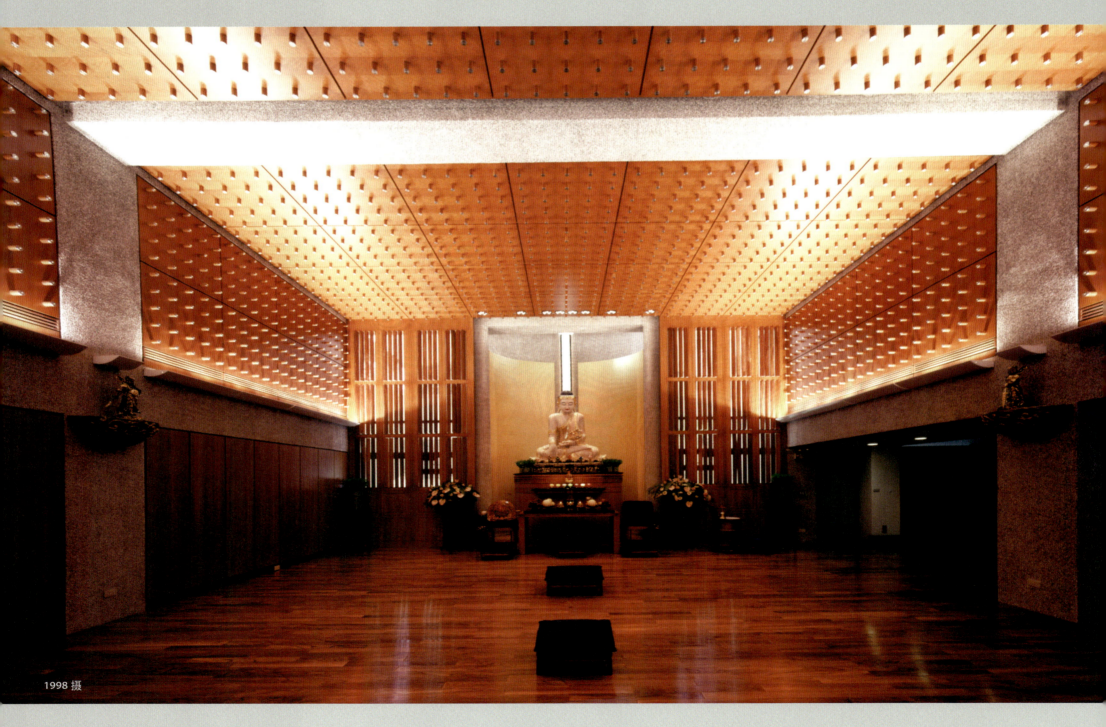

1998 摄

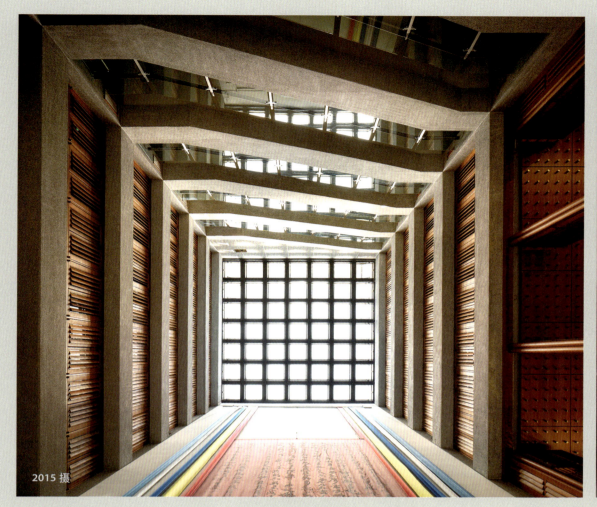

2015 摄

Skylight of the atrium courtyard
挑空中庭天窗

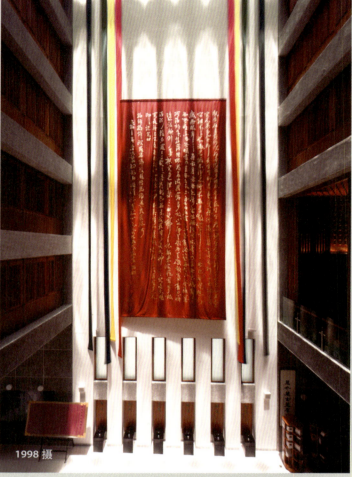

1998 摄

West wall of the atrium courtyard
挑空中庭西侧墙面

1998 摄

Details in the shrine
大殿细部

2015 摄

2015 摄

台南艺术大学音像艺术馆

TNNUA COLLEGE OF SOUND AND IMAGE ARTS

Tainan, Taiwan, China | Completion 1998
中国 台湾 台南 | 1998 年完工

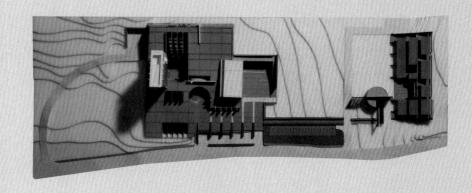

LOCATION	Tainan, Taiwan, China
CLIENT	Tainan University of the Arts (TNNUA)
FLOOR LEVELS	8 Floors, 1 Basement
BUILDING STRUCTURE	Reinforced Concrete
MATERIALS	Exposed Aggregate Finish, Textured Stucco, Sandstone
BUILDING USE	School Facilities
SITE AREA	542721 ㎡
TOTAL FLOOR AREA	10442 ㎡
DESIGN INITIATIVE	1996
COMPLETION	1998

项目位址	中国 台湾 台南
业主	台南艺术大学
楼层	地上 8 层、地下 1 层
建筑结构	钢筋混凝土
材料	洗石子、石头漆、砂岩
用途	学校
基地面积	542721 ㎡
总楼地板面积	10442 ㎡
设计起始时间	1996
完工时间	1998

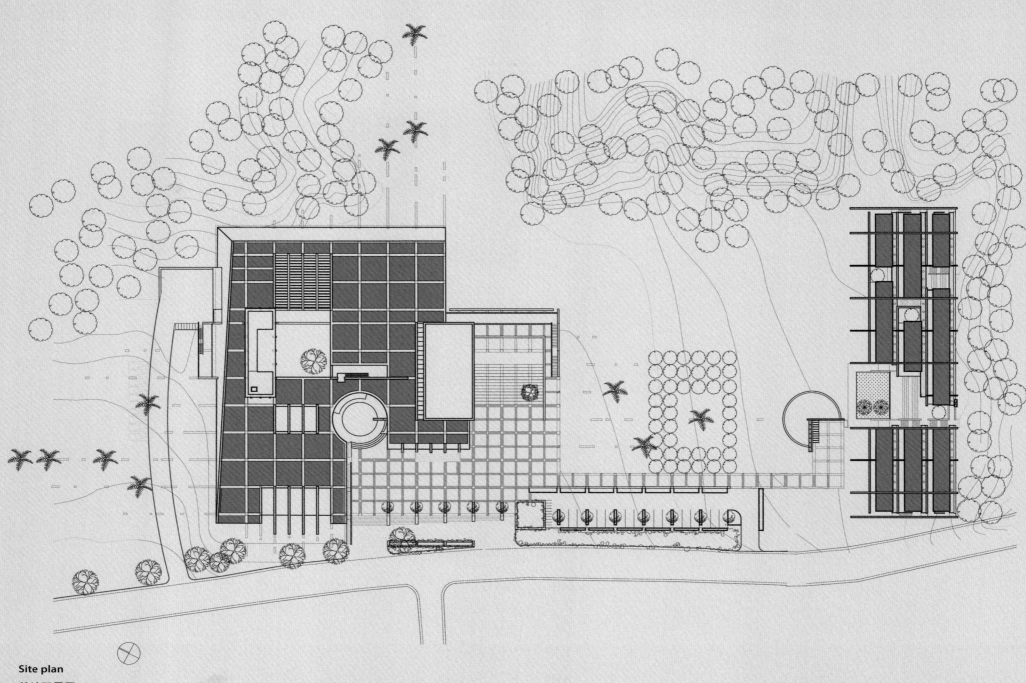

Site plan
基地配置图

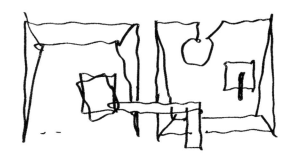

With lush green hills surrounded its three sides, the College of Sounds and Image Arts nestles its main tower and wedge-shaped podium in the valley, with only its west side open to the main campus. To further integrate into its natural surroundings, the roof of the podium, angled at a 1:6 ratio, is completely covered with soil and grass. The interior spaces underneath remain cool and comfortable as the green roof serves as a buffer to the intensive heat of southern Taiwan.

Two architecture masses protrude from the slope. The lower mass houses a movie theater, and the taller mass overlooking the nearby reservoir accommodates offices and research rooms. Courtyards of varying sizes are carved into the podium, providing light and ventilation for the lower levels and offering informal gathering spaces for students.

The dormitory complex set into the southern hill is designed as another wedge-shaped building with a 1:3 slope, providing terraces for each dormitory unit. In front of the dormitory building is a circular common room, with the same size and shape as the entry plaza to represent a cut out of the main building.

音像艺术学院的主楼与楔形墩座隐身于山谷之中，三侧环绕着蓊郁的山丘，只有西侧直接迎向主校区。呈 1 : 6 角度偏斜的墩座屋顶以土壤及绿草复盖，完全融入四周的自然环境。绿屋顶挡住了台湾南部的酷热，使下方的室内空间能够保持凉爽舒适。

两栋建筑主量体突出于山坡上，低矮的建筑量体内设有一间电影院，高楼层量体可俯瞰附近的水库，内部为研究室及办公室。墩座内崁嵌大小不一的室外中庭，除可为低楼层提供良好的采光和通风，也可作为学生非正式的聚会空间。

配置于南方山丘的宿舍大楼是另一栋 1 : 3 斜率的楔形建筑，每间宿舍都设有阳台。宿舍大楼前方的圆形交谊厅与主建筑的入口广场形状、大小完全相同，虚实量体两相呼应。

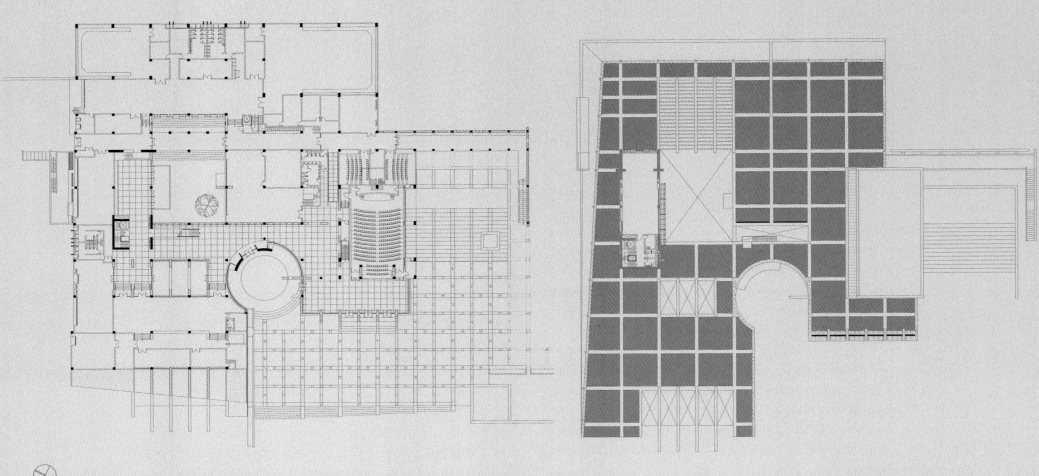

Ground floor plan
地面层平面图

Fourth floor plan
四层平面图

South facade
南向立面

Tower view from south
塔楼南侧立面

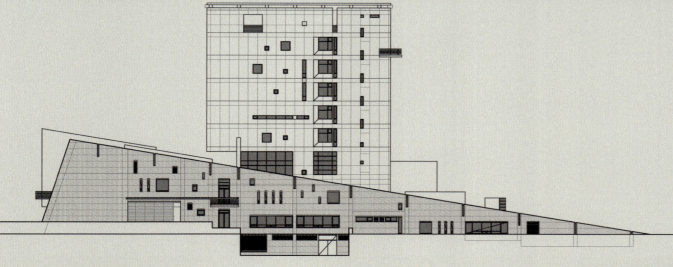

Northwest elevation
西北向立面图

Courtyard and exterior passageway

中庭及室外连通道

SHIH CHIEN UNIVERSITY GYMNASIUM AND LIBRARY

实践大学体育馆＋图书馆 Taipei, Taiwan, China | Completion 2009
中国 台湾 台北 | 2009 年完工

Third floor terrace at the library
图书馆三楼平台

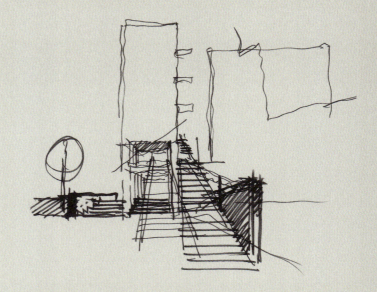

LOCATION	Taipei, Taiwan, China
CLIENT	Shih Chien University
FLOOR LEVELS	5 Floors, 2 Basements
BUILDING STRUCTURE	Reinforced Concrete, Steel Structure Roof (Gym)
MATERIALS	Architectural Concrete, Aluminum Louver, Clear Glass, Stainless Steel Mesh
BUILDING USE	School Facilities
SITE AREA	42039 ㎡
TOTAL FLOOR AREA	Gymnasium 14818 ㎡
	Library 12031 ㎡
DESIGN INITIATIVE	2003
COMPLETION	2009

项目位址	中国 台湾 台北
业主	实践大学
楼层	地上 5 层、地下 2 层
建筑结构	钢筋混凝土、钢骨结构屋顶（体育馆）
材料	清水混凝土、铝挤型百叶、清玻璃、不锈钢编织网
用途	学校
基地面积	42039 ㎡
总楼地板面积	体育馆 14818 ㎡
	图书馆 12031 ㎡
设计起始时间	2003
完工时间	2009

Model, aerial view
模型，鸟瞰

Continuing the success of the College of Design building also designed by Artech, the Shih Chien University decided to bring the two most important public spaces on the campus — the gymnasium and the library — to a grander scale and to form a green new quad as the second gateway for the campus. The two buildings are joined together by a subterranean passageway that serves as a cultural/conference center. Though one is a space of silence while the other is a space for dynamic activities, their spatial fluidity and openness are both the catalysts to promote interpersonal interaction and eliminate barriers.

The building materials used were chosen to blend in with the overall style of the campus. For the gymnasium, stainless steel mesh and architectural concrete demarcate the hollow and the solid, differentiating between the attributes of the large and small gym spaces. The mesh and vertical louvers of the library's façade reflect the materials used for the gymnasium, while acting as sun shading device. In the central quad there are irregularly shaped light wells extending down to the underground level, providing improved lighting for more versatile use.

实践大学延续设计教学大楼的理念与追求，将校园最重要的两个公共空间"体育馆"与"图书馆"带向更宏观的校园公共建筑思维。体育馆作为西侧入口迎面配置，东侧为图书馆，两栋重要建筑合抱共同围塑绿地空间，重新创造校园核心，并于地下设置艺文及会议中心以通道串联，让新的广场成为校园的第二门户意象。两栋建筑内部活动动静有别，但皆以空间的流动与开放性来促成校园人际间的联系，成为消除界线的触媒。

建筑材料延续校区整体风格；体育馆量体以不锈钢编织网与清水混凝土作为虚体与实体的上、下交错，区分大、小跨距的运动空间属性，通过外墙编织网，由室外可见室内人们活动的光影；图书馆以立面格栅呼应体育馆的材料应用，兼顾建筑节能的手法，让阅读的温度与光线更加自然。中庭绿地设置不规则造型的地下室采光井，使地下空间更加明亮，增加使用的多变性。

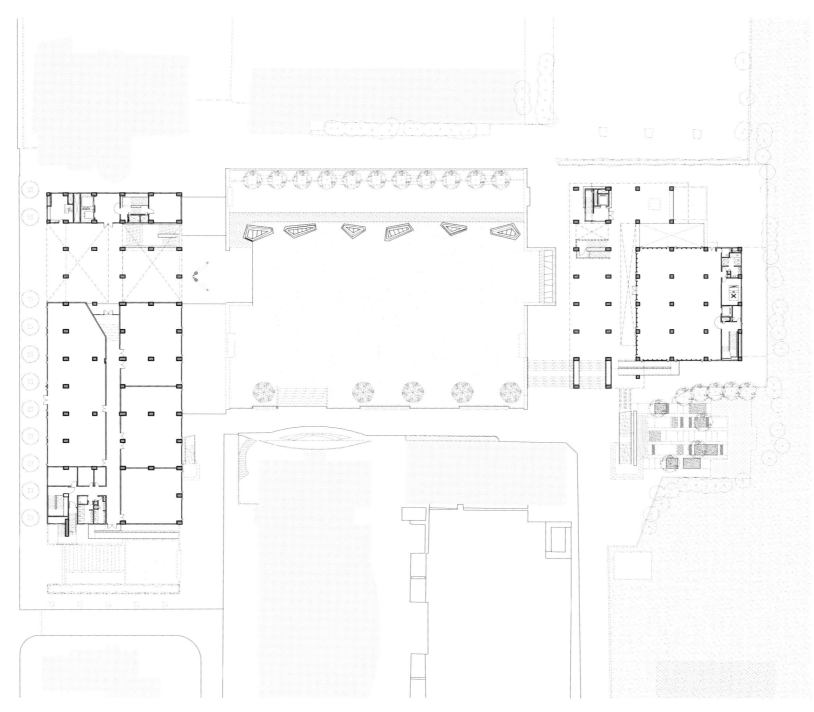

Ground floor plan
地面层平面图

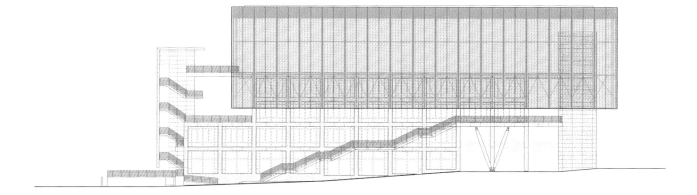

Gymnasium east elevation
体育馆东向立面图

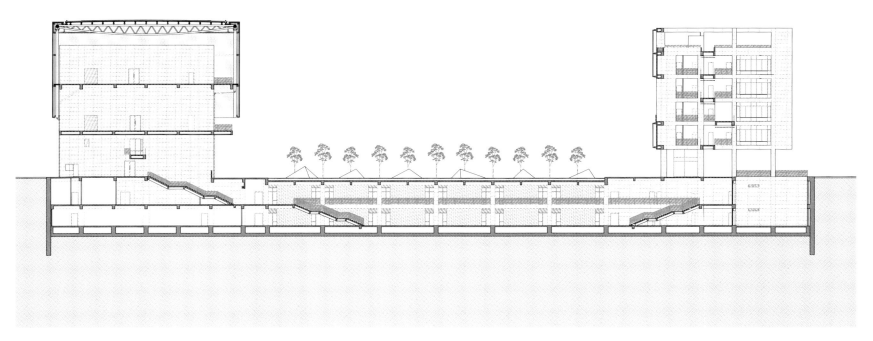

Cross section
全区横向剖面图

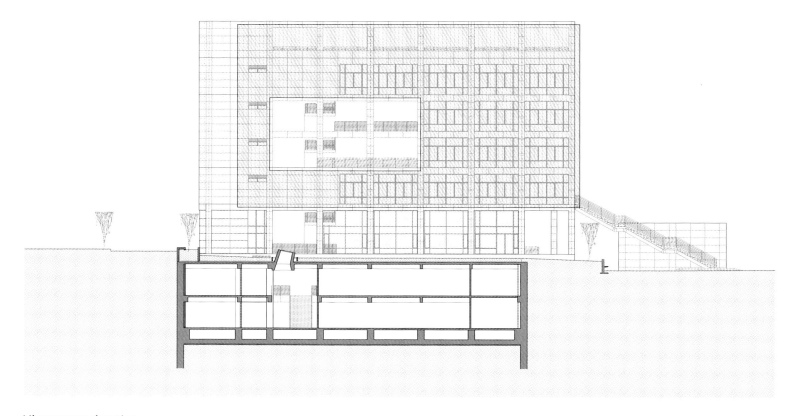

Library west elevation
图书馆西向立面图

Library ground floor façade detail
图书馆一层局部立面

Details of the architectural concrete wall,
louvers, and the exterior stairway
清水混凝土外墙、铝百叶及室外梯细部

Exterior stairway to the second floor of library
通往图书馆二楼室外梯

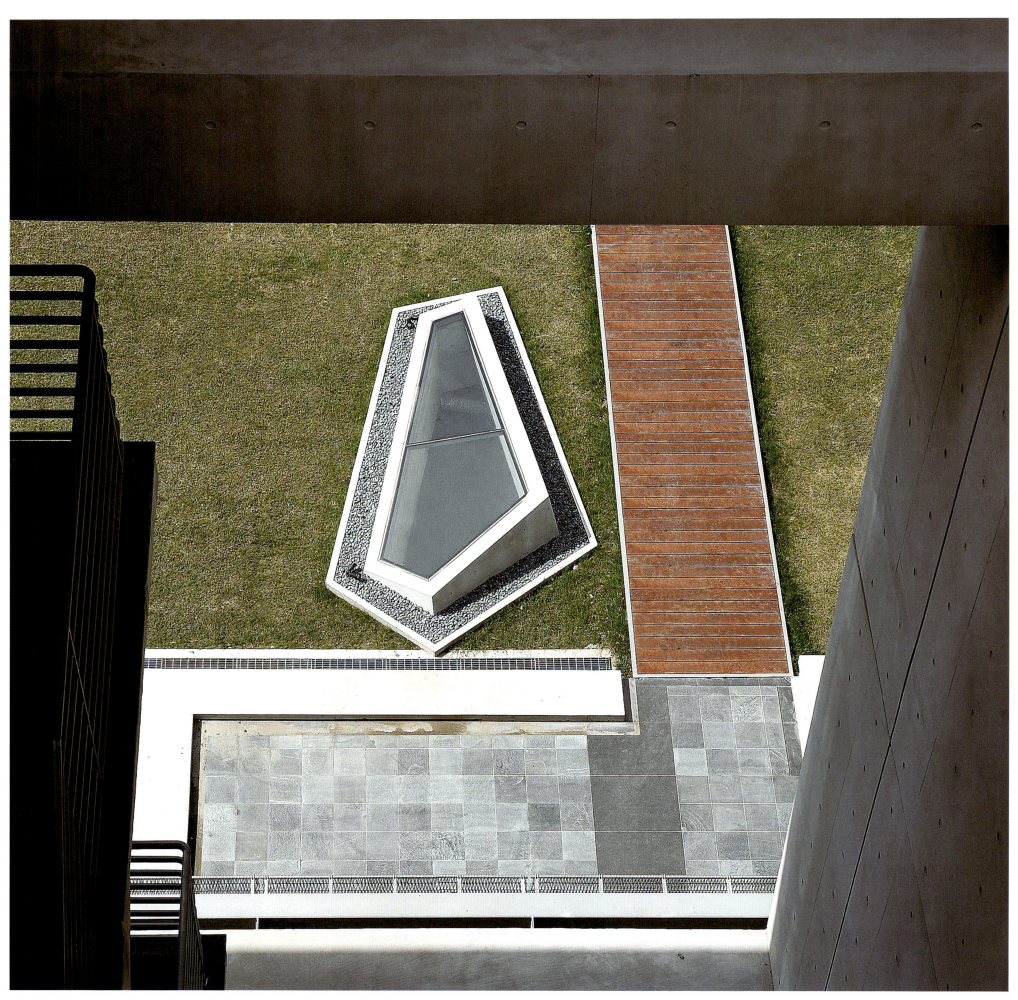

Skylights and the green quad
绿地广场及天窗

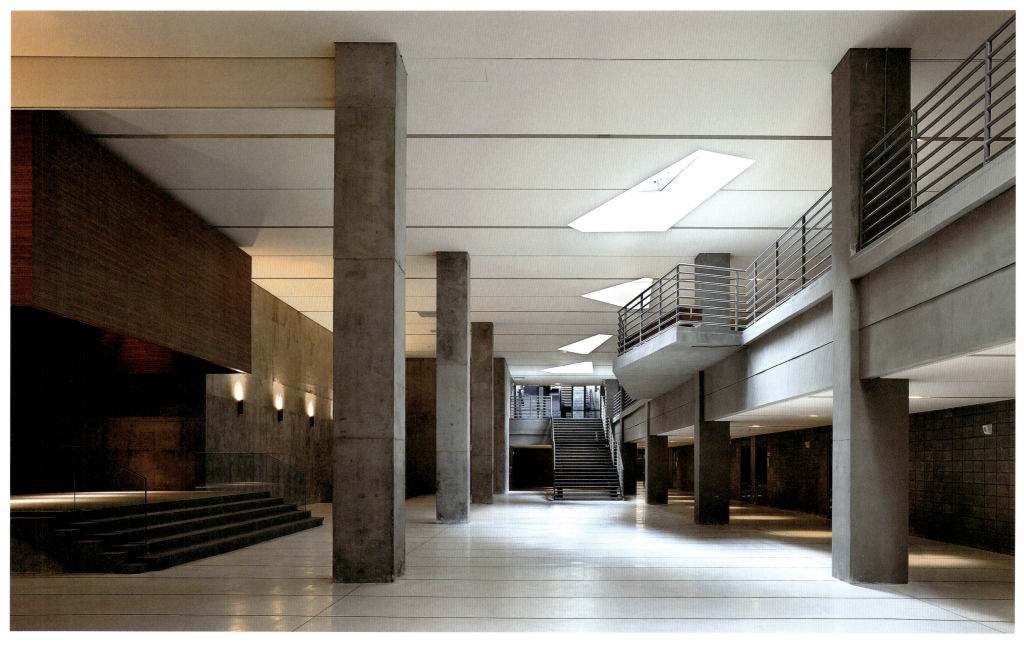

Underground hall
地下通廊

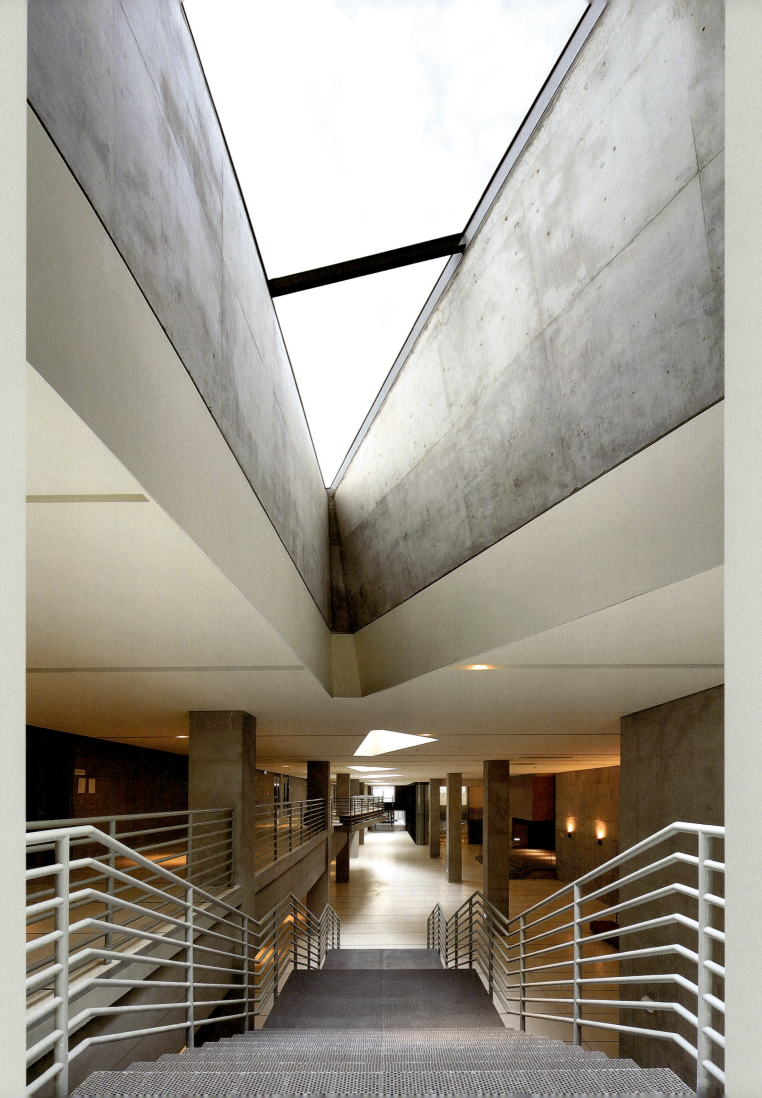

View to gymnasium from the green quad
从绿地广场望体育馆立面

The basketball court
体育馆篮球场

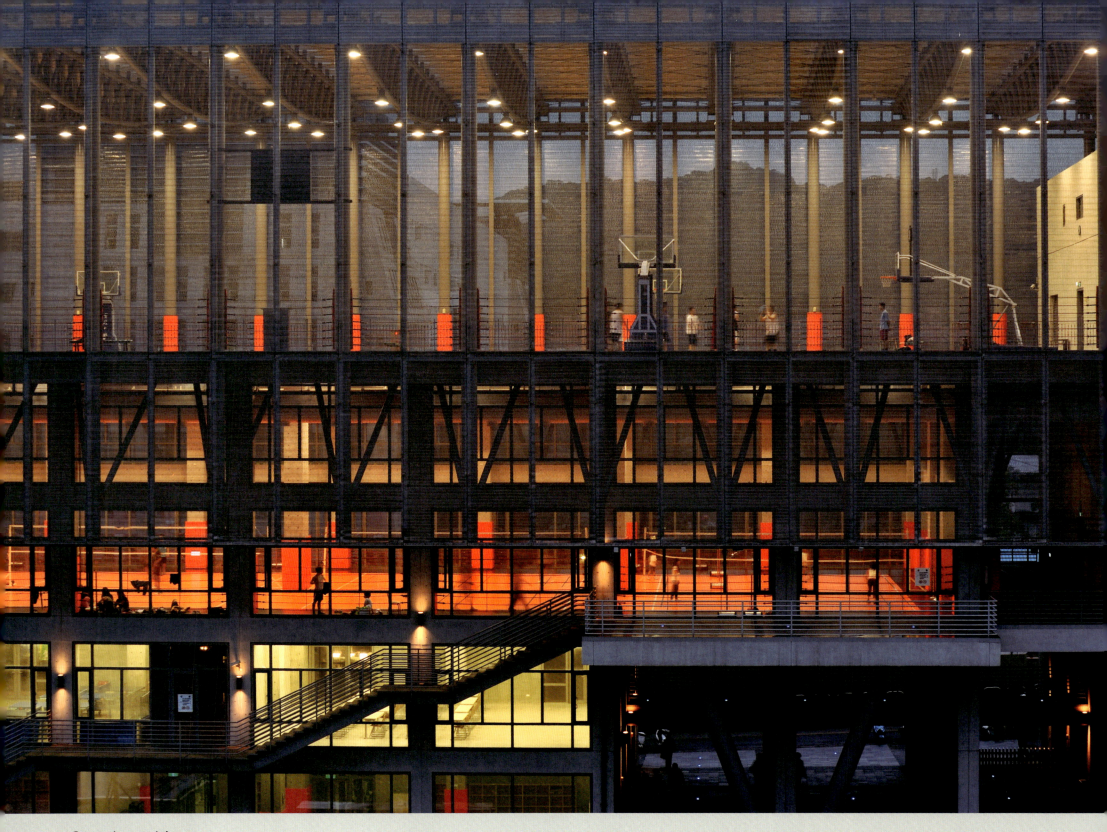

Gymnasium at night
体育馆夜景

Library ground floor reading room at night
图书馆一楼阅览室夜景

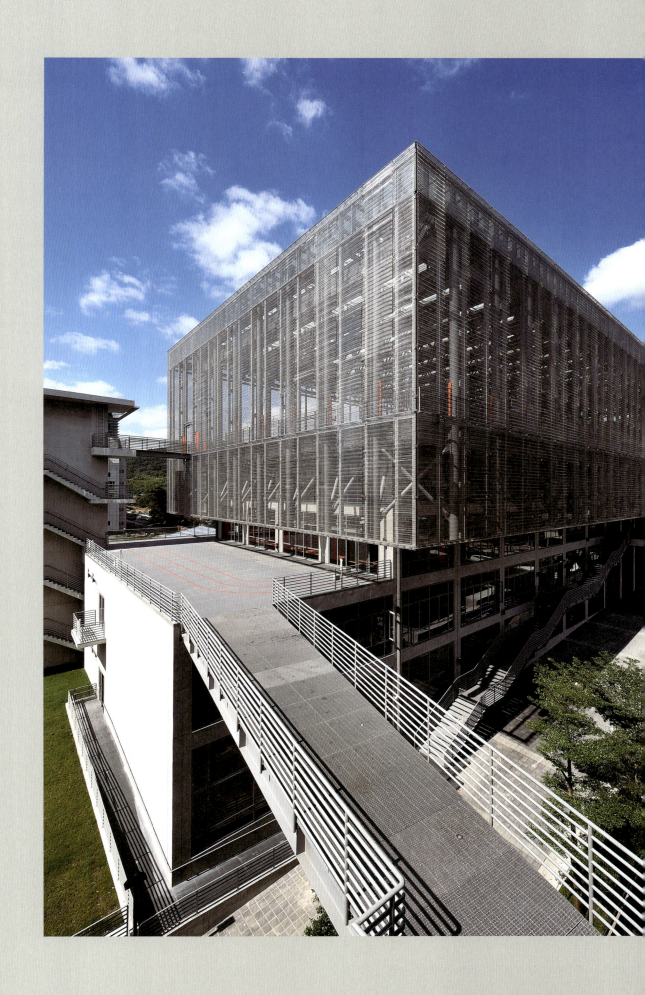

CHIAO TUNG UNIVERSITY GUEST HOUSE

交通大学招待所

Hsinchu, Taiwan, China | Completion 2008

中国 台湾 新竹 | 2008 年完工

LOCATION	Hsinchu, Taiwan, China
CLIENT	Chiao Tung University
FLOOR LEVELS	3 Floors, 1 Basement
BUILDING STRUCTURE	Reinforced Concrete
MATERIALS	Teak Wood Facade Panel, Alum num Panel, Exposed Aggregate Finish
BUILDING USE	Guest House
SITE AREA	2450 ㎡
TOTAL FLOOR AREA	3099 ㎡
DESIGN INITIATIVE	2003
COMPLETION	2008

项目位址	中国 台湾 新竹
业主	交通大学
楼层	地上 3 层、地下 1 层
建筑结构	钢筋混凝土
材料	柚木外饰板、烤漆铝板、抿石子
用途	招待所
基地面积	2450 ㎡
总楼地板面积	3099 ㎡
设计起始时间	2003
完工时间	2008

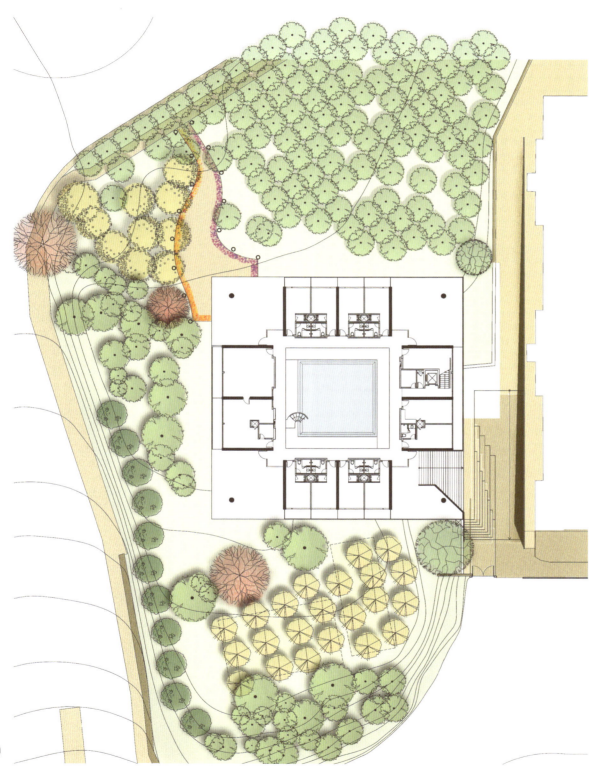

Ground floor plan
地面层平面图

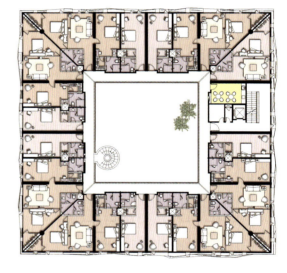

Fourth floor plan
四层平面图

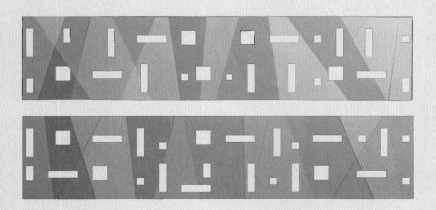

Facade concept
立面概念

Located on a slope adjacent to the north gate of the main campus, the university Guest House sits quietly in the midst of the pinewoods, camouflaged as a simple floating box with random openings, folded surfaces and a teak facade.

Visitors enter from underneath the wooden box to encounter a tranquil courtyard carved out from the cube. The courtyard is white in color, with a square pond in the middle and bamboo landscape surrounding its edges. It reflects the sky and transmits serenity into this natural setting.

In the guestrooms, strong colors are applied on the ceiling, walls and floors for the private quarters. Personal-scaled windows that are placed strategically for guests to view the pinewoods, either vertical or horizontal, connect the guests to the natural scenery outside.

此招待所位于校区北大门旁的坡地上，柚木外观的折板立面，装点着随性的开窗，安静地坐落在松木林间，仿佛一只漂浮的木盒子。

访客从木盒下方进入时，首先映入眼帘的是一座宁静的中庭。一池方形水塘位于白色庭院中央，四周包围了竹林景观，倒映着上方的天空，也为此建筑物增添了静谧感。

客房内的天花板、墙壁和地板均采用强烈色彩。依照人体尺度的开窗设计饶富随性的趣味，或垂直或水平，访客能以各种方向观赏窗外的松木林。

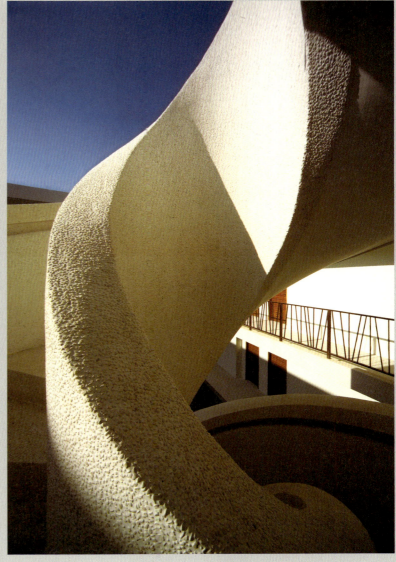

水月道场

Taipei, Taiwan, China | Completion 2012
中国 台湾 台北 | 2012 年完工

WATER-MOON MONASTERY

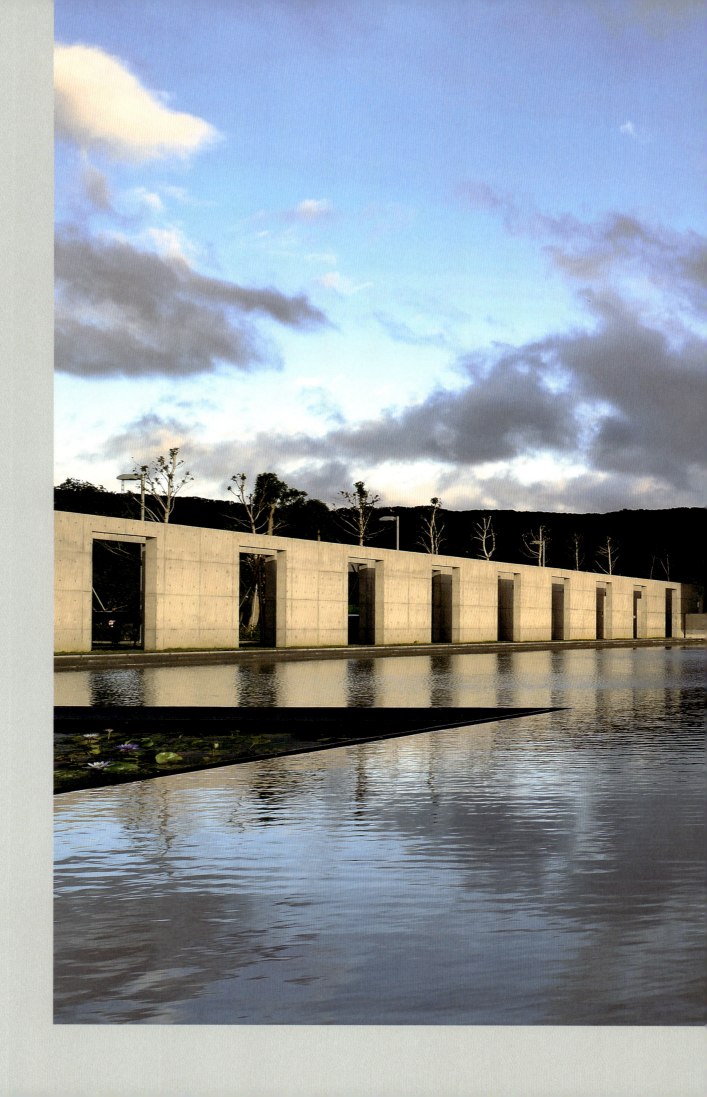

LOCATION   Taipei, Taiwan, China
CLIENT   Dharma Drum Mountain Nung Chan Monastery
FLOOR LEVELS   2 Floors, 1 Basement
BUILDING STRUCTURE Steel Structure, Reinforced Concrete
MATERIALS   Architectural Concrete, Teakwood, Limestone, Glass
BUILDING USE   Religious
SITE AREA   27936 ㎡
TOTAL FLOOR AREA   8056 ㎡
DESIGN INITIATIVE   2006
COMPLETION   2012

项目位址   中国 台湾 台北
业主   法鼓山农禅寺
楼层   地上 2 层、地下 1 层
建筑结构   钢骨结构、钢筋混凝土
材料   清水混凝土、缅甸柚木、莱姆石、玻璃
用途   宗教
基地面积   27936 ㎡
总楼地板面积   8056 ㎡
设计起始时间   2006
完工时间   2012

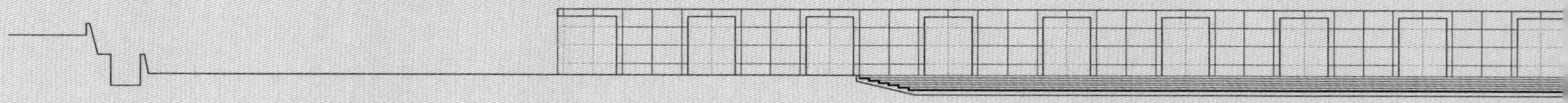

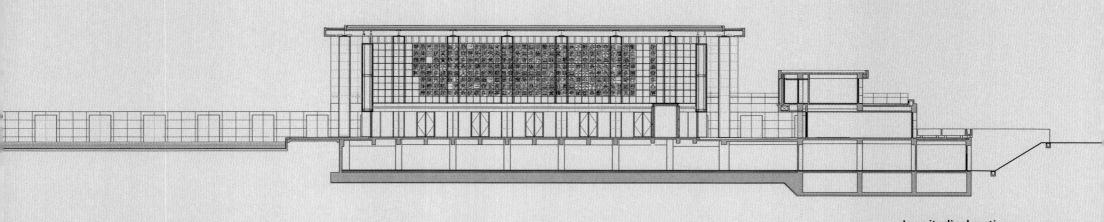

Longitudinal section
纵向剖面图

When asked of what his vision for the future temple would be, Master Sheng Yen, the founder of the Monastery and Dharma Drum Buddhist Group, answered that he "sees" the temple in his meditation dhyana, "It is a Flower in Space, Moon in Water," he said. "Let's name it the Water-Moon Monastery".

Thus began the Water-Moon Monastery. Situated on the vast Guandu plain, facing the Keelung River and with the Datun Mountain as its backdrop, the design takes advantage of its natural surroundings and strives to build a tranquil spiritual place.

After passing through two walls of different heights that serve as buffer from the expressway outside, upon entering the temple, visitors face the view of the shrine that sits at the far end of an 80-meter long lotus pond. The reflections on the pond of the over-sized colonnades and the flowing golden drapes in between create a scene of illusory quality. Using architectural concrete as the main material, the design reduces the color and form to a minimum, conveying the spirit of Zen Buddhism. The lower part of the shrine is transparent, giving it an impression of its upper wooden "box" being suspended in the air.

On the west side of the shrine, a massive wooden wall is carved with the famous "Heart Sutra" in Chinese characters. As the lights shine through the carved-out characters, the space is infused with an aura of culture and spirituality. Outside the long corridor, the characters of the "Vajracchedika Prajnaparamita Sutra" are cast void on the prefabricated GRC panels, providing additional religious meaning while functioning as sunshades. When the scripture is imprinted onto the interior surface by the sunlight, it is as if the Buddha's teaching, in an unspoken manner, is revealed.

当法鼓山佛教学院创办人圣严法师被问及对未来农禅寺的想法时，他表示曾在入定时"看到"寺庙的样貌，有如"空中花，水中月"。于是他说："我们就取名为水月道场吧。"

坐落于广大关渡平原的农禅寺水月道场于焉诞生。面向基隆河，背倚大屯山；利用这优美灵秀的环境，营造一处清雅幽静的宗教空间。

访客一开始先穿越两面高度不同的墙，作为与外头高速道路之间的缓冲；一进入道场，即能看到远方的主讲堂，静静伫立于 80 米长的荷花池中。超大柱廊在池中的倒影，伴随着飞扬其间的金色帘幔，自成一虚幻雅致风光。主材料利用建筑混凝土，设计上尽量摒除华丽色彩与装饰，意图传达简朴的禅佛况味。大厅的下半部的透明无柱设计，为上半部的木头盒子带来空悬于上的缥缈幻象。

大厅西面厚实的木墙上刻着"心经"，当阳光透过镂刻的经文洒进，空间瞬间充满修养灵性氛围。长廊外的"金刚经"则是在玻璃纤维混凝土预制板上镂空的字，既能遮阳，更增添了宗教意义。阳光洒落时，穿透经文，铭刻到内部墙面，仿佛为众人揭示无声之法。

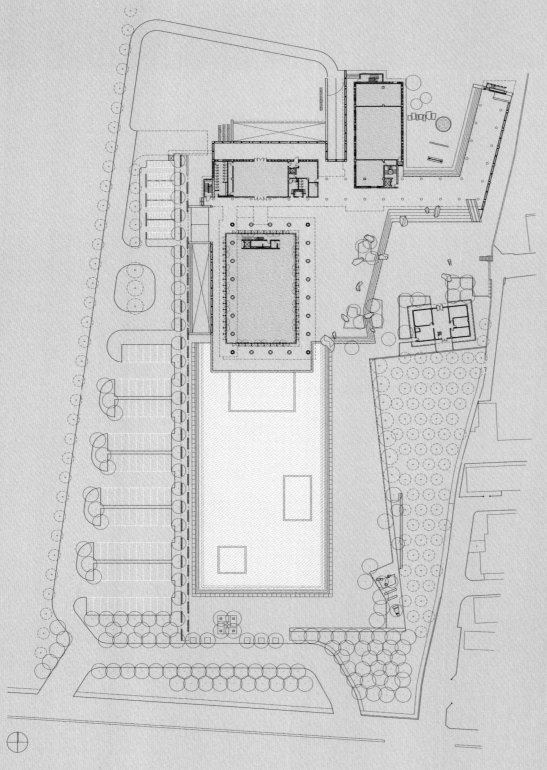

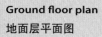

Ground floor plan
地面层平面图

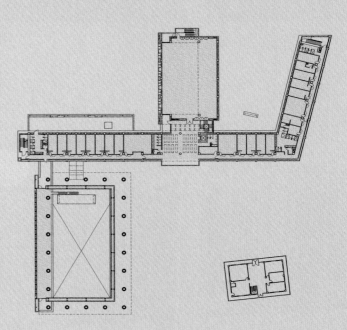

Second floor plan
二层平面图

Model, courtyard

模型，中庭

法鼓文化提供．邓博仁摄

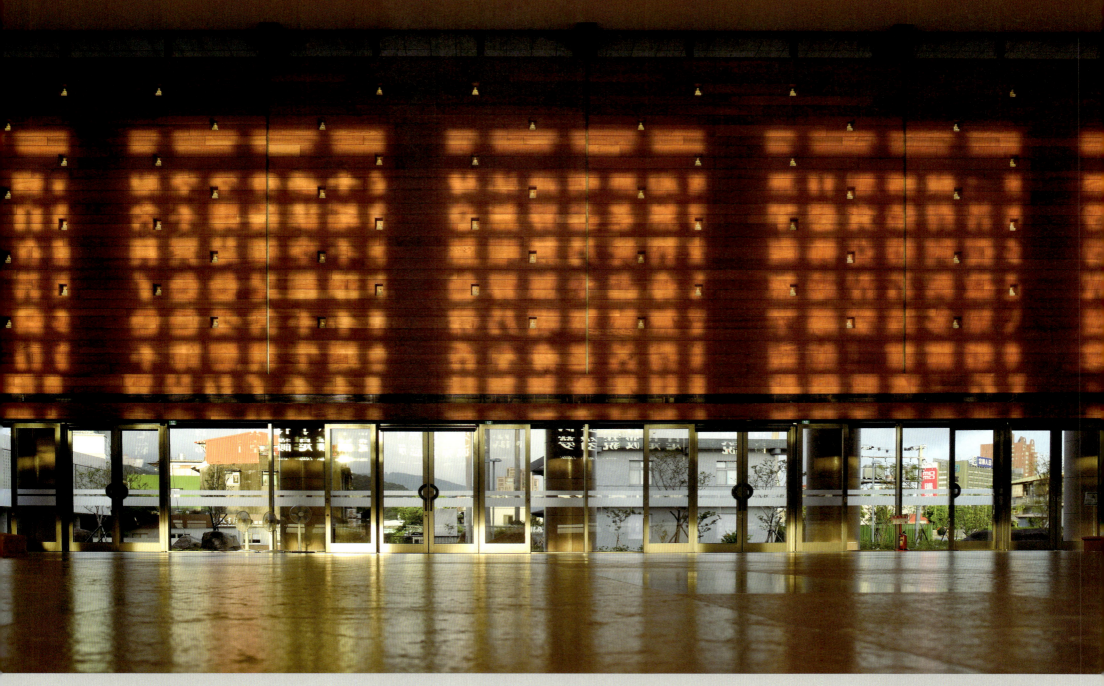

Interior of the shrine
大殿室内

Sunlight through the Heart Sutra wall
"心经"墙的光影

法鼓文化提供，邓博仁摄

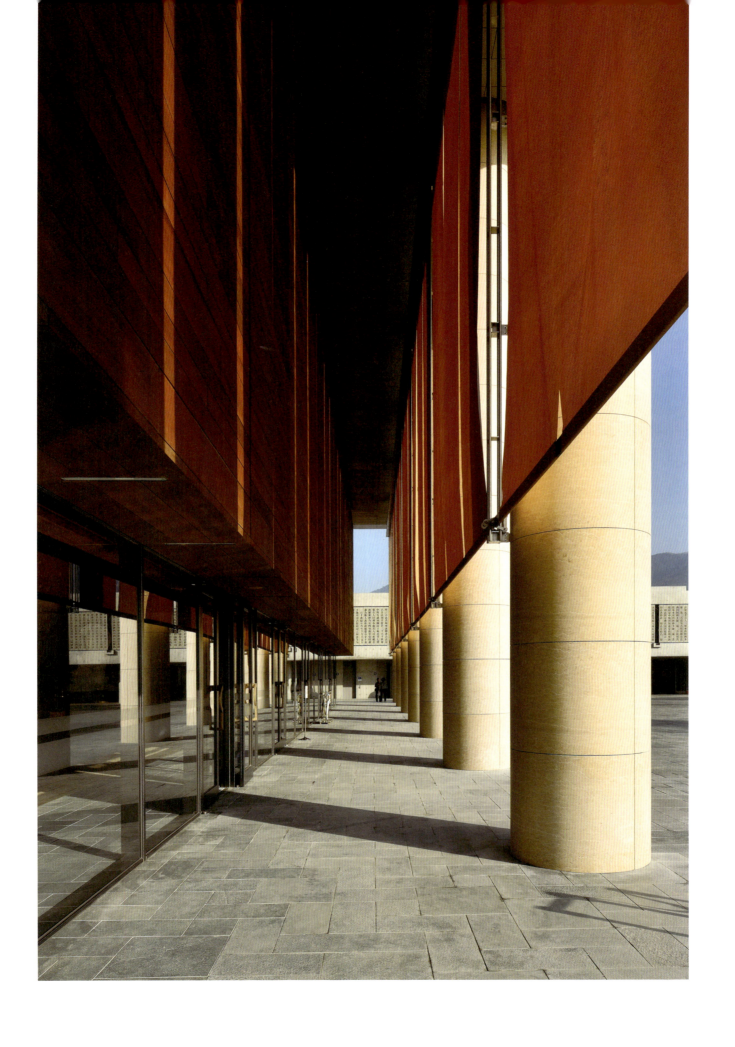

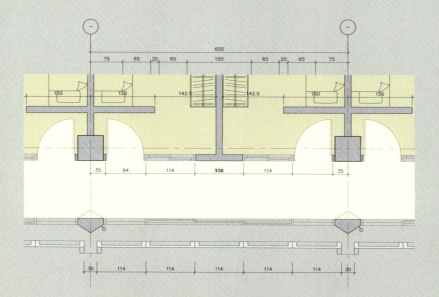

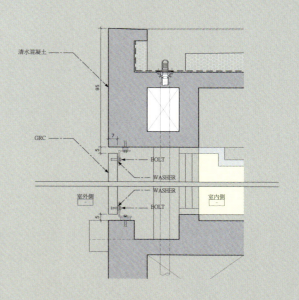

Details of the GRC Diamond Sutra wall
玻璃纤维混凝土预制板"金刚经"墙细部

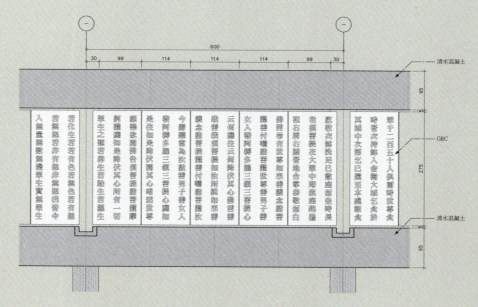

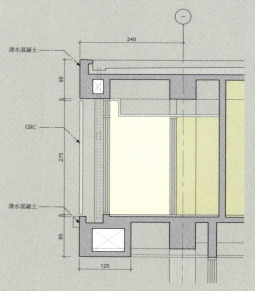

法鼓文化提供，李东阳摄

Meditation hall foyer

二楼禅堂前厅

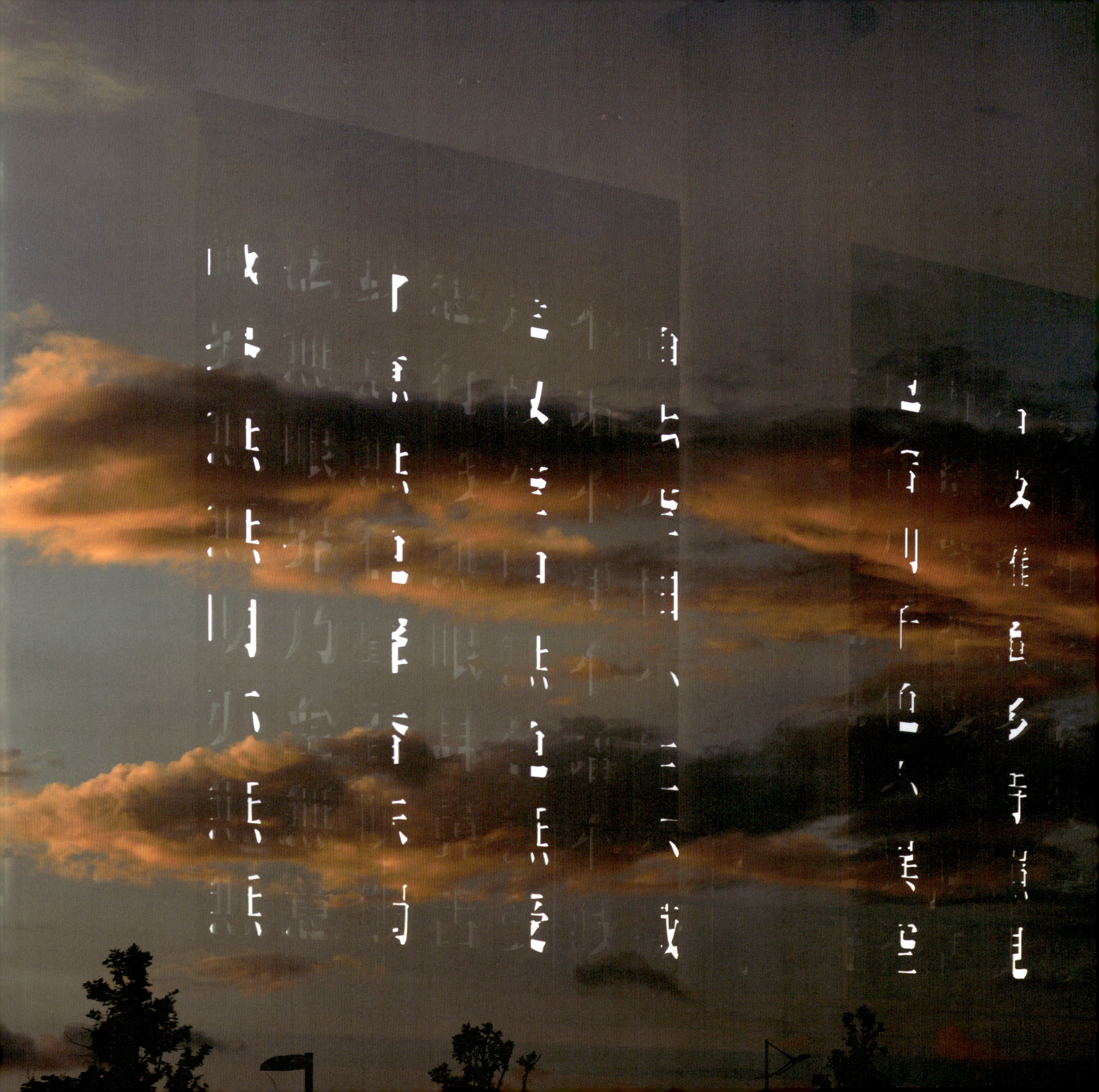

法鼓文理学院

DHARMA DRUM INSTITUTE OF LIBERAL ARTS

New Taipei, Taiwan, China | Completion 2015
中国 台湾 新北 | 2015 年完工

LOCATION	New Taipei, Taiwan, China
CLIENT	Dharma Drum Institute of Liberal Arts (DILA)
FLOOR LEVELS	6 Floors, 2 Basements
BUILDING STRUCTURE	Reinforced Concrete
MATERIALS	Architectural Concrete, Clear Glass, Aluminum Panel, Titanium Zinc Plank, Exposed Aggregate Finish, Wood
BUILDING USE	School Facilities
SITE AREA	249113 ㎡
TOTAL FLOOR AREA	Academic/Administration Building 23050 ㎡
	Dormitory 12155 ㎡
	Sports Complex 6833 ㎡
DESIGN INITIATIVE	1998
COMPLETION	2015

项目位址	中国 台湾 新北
业主	法鼓文理学院
楼层	地上 6 层、地下 2 层
建筑结构	钢筋混凝土
材料	清水混凝土、清玻璃、铝板、钛锌板、抿石子、实木
用途	学校
基地面积	249113 ㎡
总楼地板面积	行政及教学大楼 23050 ㎡
	禅悦书苑 12155 ㎡
	体育馆 6833 ㎡
设计起始时间	1998
完工时间	2015

This project is located on a gentle wooded slope. The school focuses on harmony with nature and ecologically sustainable growth; therefore, preserving the original site terrain is of utmost criteria. The buildings grow out from the earth, in full harmony with nature. The buildings are mostly low-rise structures with minimal excavation needed. The design of a total of 1590 exterior and interior steps allows access to different elevations on the hillside while eliminating any visible retaining walls. Three main buildings are positioned according to site conditions: the Academic/Administration Building, the Chan House, and the Sports Complex. Platforms, corridors and bridges link all the spaces together, providing free access, expanding the activity areas, and highlighting the dramatic qualities of the adjoined spaces. The overall style is imbued with a sense of age, so upon completion the buildings show no unexpected novel elements, and will retain its features even after many decades.

The Academic/Administration Building exhibits its timelessness with architectural concrete finishes, and its rooftop greenery helps make it one with nature. The long stairway follows the contours is the primary focus for campus life. With a predominant rainy weather, many corridors, balconies and other semi-outdoor spaces are provided for student's activities. They also enhance visual vistas and create many both interesting framing views.

The Chan House is a student dormitory. The buildings are scattered on the mountainside, configured so that each building's rooftop is at a different level, increasing the usability of the roof platforms. Pebbledash façades create a simple, clean style, and the atrium gardens contain pools that attract birds, and acts as a physical separation between men's and women's quarters.

The Sports Complex is built with architectural concrete finishes, composed of one square and one linear volumes. The square volume has a 7-meter high ceiling, suitable for any sports activity venue. The linear volume is made up of irregular convex and concave box-like spaces for an intriguing visual effect. Skylights let in sunlight for interesting light and shade, and the design takes advantage of the sloping terrain of the valley to channel fresh air into the interior spaces. The natural lighting and ventilation make it an excellent place for activities and contemplation.

本案坐落于广大的林地山坡，校方着重与自然和谐共存以及生态永续发展，校区开发尽可能维持环境的本来面目。建筑物有如自大地生长而出的有机体，与大自然融谐无碍。本案分期建造，设计上多以低矮的多层次建筑为主，减少开挖，以 1590 个台阶消化山坡地高差的挡土墙。因地制宜配置三栋主要建筑物：行政及教学大楼、禅悦书苑、体育馆。建筑物利用平台、回廊与通桥串联每个空间，出入自由，延伸活动场域，展现连贯的空间戏剧。整体风格具有岁月的古感，使建筑物落成之初，不显新颖突兀，而经过数十年仍能保有外观特色。

行政及教学大楼以清水混凝土展现隽永风格，平屋顶植栽绿化与自然合一。依山势而建的大型阶梯成为校园生活的主要场景；为因应多雨气候，大量留设虚空间提供学生更多活动场所，通透的视野与格栅的搭配，与走廊、阳台等半户外空间交错，室内户外视角穿透，相互借景，制造框景效果。

禅悦书苑是学生宿舍，依山势错落配置，使各栋高度相差一层楼，增加平台屋顶利用，抿石子立面营造简单纯粹风格。中庭花园有诱鸟池，具有听溪赏水的雅致，并区隔出男女学生住宿空间。

体育馆以清水混凝土建造，一长一方的量体并置，方量体挑高 7m，创造出宽阔活动场地，长量体错落设置着内缩或外推盒状空间，形成有趣的框景。建筑物以天窗采光，光影交错洒落，利用地形引导坡地与溪谷气流进入室内空间，在天光与徐风中有着暂留与沉思的场所。

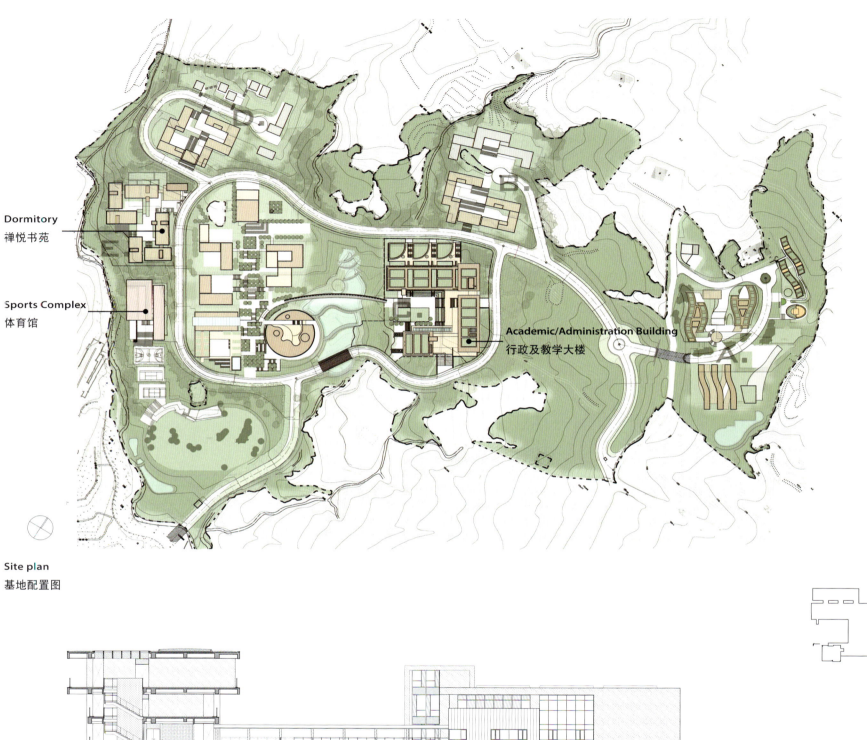

Dormitory
禅悦书苑

Sports Complex
体育馆

Academic/Administration Building
行政及教学大楼

Site plan
基地配置图

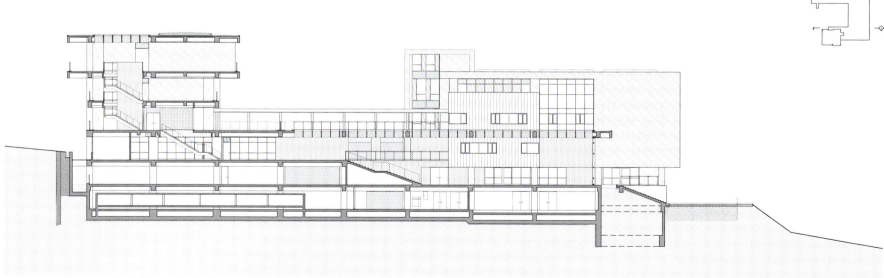

East-west section (Academic/Administration Building)
行政及教学大楼东西向剖面图

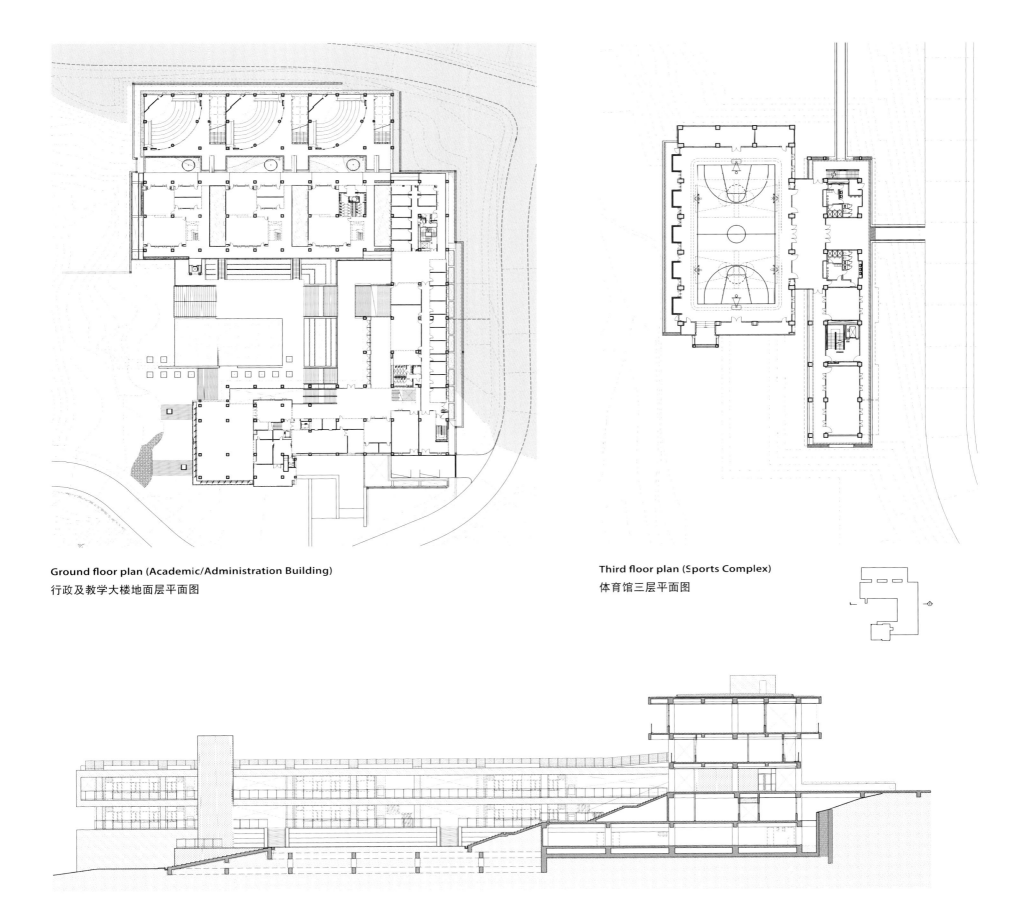

Ground floor plan (Academic/Administration Building)
行政及教学大楼地面层平面图

Third floor plan (Sports Complex)
体育馆三层平面图

East-west section (Academic/Administration Building)
行政及教学大楼东西向剖面图

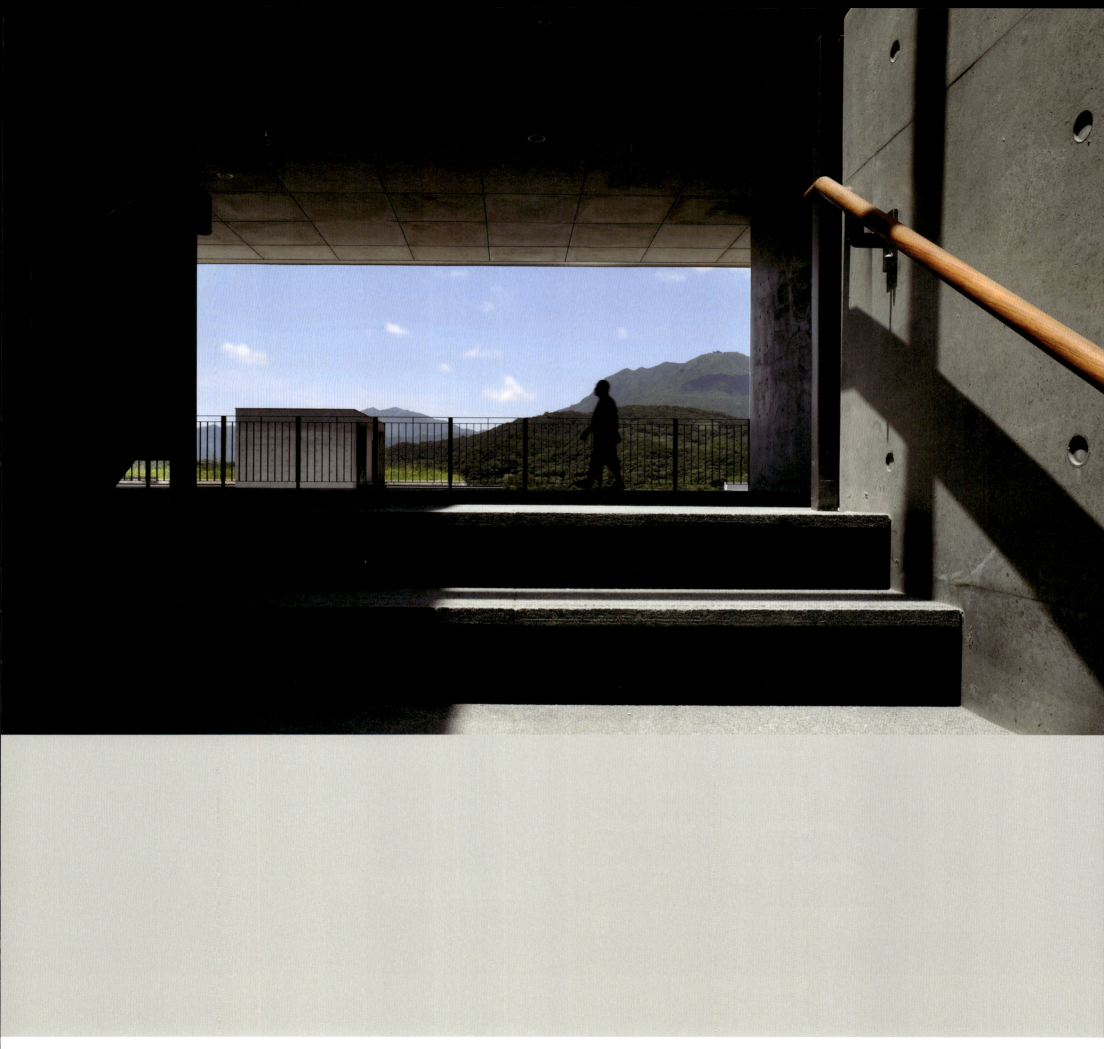

Facade Details
外观细部

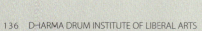

South facade of the sports complex
体育馆南向立面

佛手湖轩

Nanjing, China

中国 南京

THE DRAPE
HOUSE

LOCATION	Nanjing, China
CLIENT	Nanjing Foshou Lake Architecture & Art Development Ltd.
FLOOR LEVELS	2 Floors, 1 Basement
BUILDING STRUCTURE	Steel Structure, Reinforced Concrete
MATERIALS	Fabric, Teakwood
BUILDING USE	Vacation House
SITE AREA	10152 ㎡
TOTAL FLOOR AREA	640 ㎡
DESIGN INITIATIVE	2003

项目位址	中国 南京
业主	南京佛手湖建筑艺术发展有限公司
楼层	地上 2 层、地下 1 层
建筑结构	钢骨结构、钢筋混凝土
材料	布、柚木
用途	度假屋
基地面积	10152 ㎡
总楼地板面积	640 ㎡
设计起始时间	2003

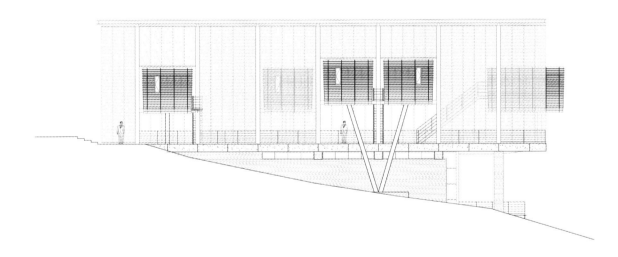

West elevation
西向立面图

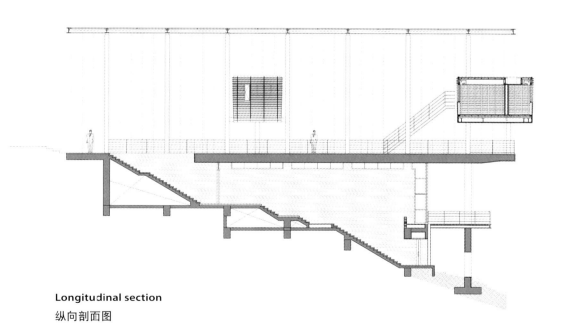

Longitudinal section
纵向剖面图

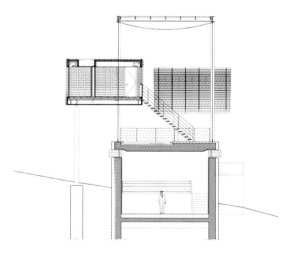

Cross section
横向剖面图

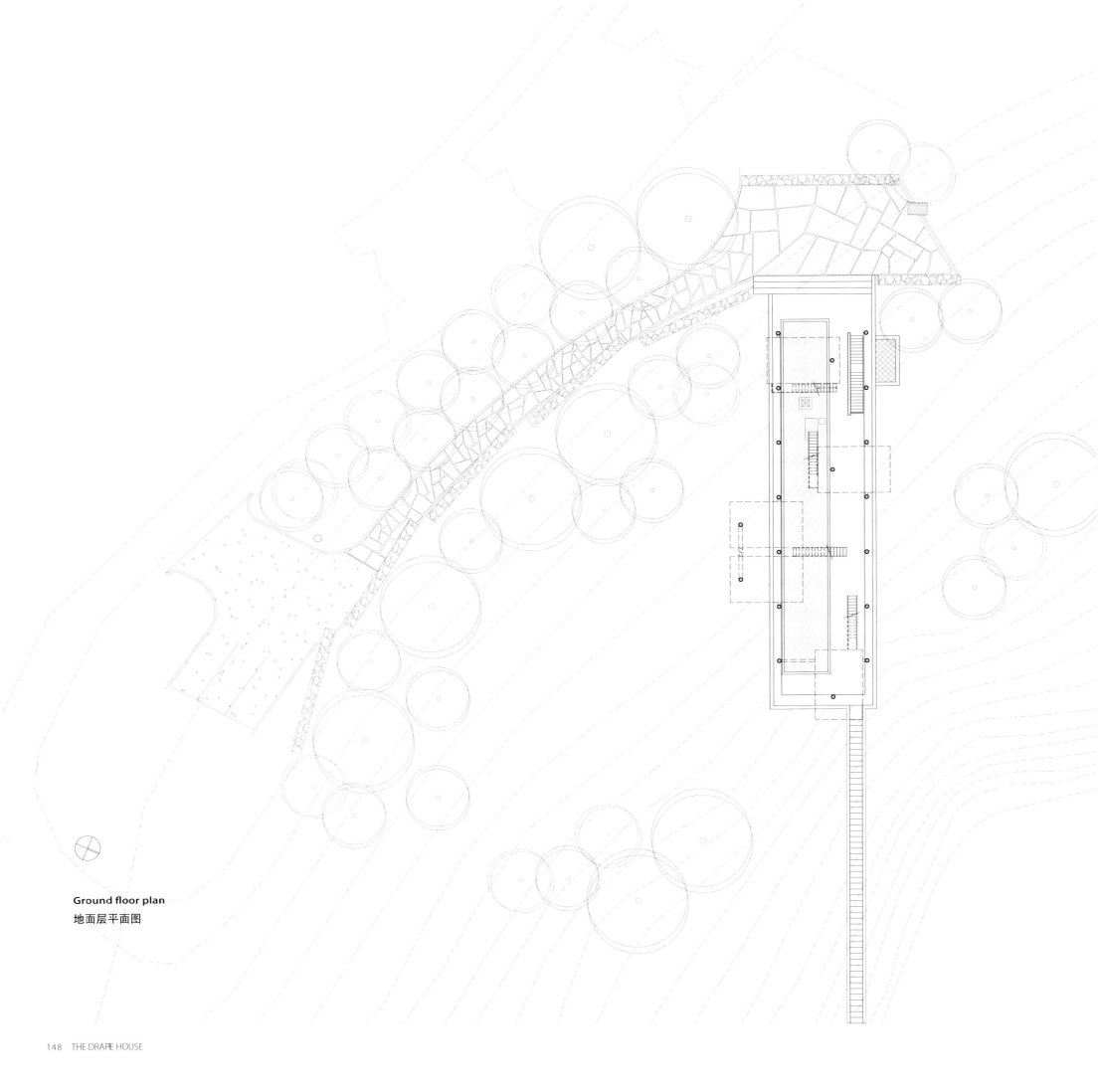

Ground floor plan
地面层平面图

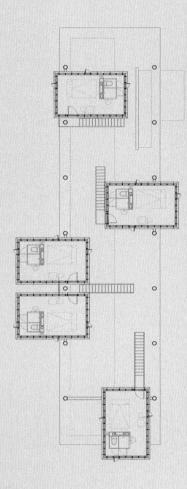

Upper floor plan
上层平面图

Twenty architects from around the world were invited to each design a five-room vacation house on the shore of the Fo Sho Lake in Nanjing, China. Our team proposed a "House of Drapes" by the lake. Emulating a traditional dyehouse with hanging fabrics, the swaying draperies obscure the frame of the building, creating indefinite forms that are constantly changing as the wind blows.

The estate is not fully visible at first glance when approached. The house slowly reveals itself as the visitor strolls along a winding path, passing a swing, a rusted animal cage, a juniper-burning stove to finally catch glimpses of the building through the trees. A view of the lake unfolds at the end of the path as the visitor reaches the lakefront house. At the ground floor entrance, five box-shaped wooden structures are suspended in the air like tree houses, held by a steel frame. Maroon drapes are hung on the sides of the boxes. Stairs that are partially visible from the exterior connect these wooden compartments independently. The elongated pool on the ground, reflecting the house, enhances the connection between the structure and the lake, which serves as the axis of this linear structure. The kitchen, dining, and living spaces are hidden from view, in the slopes of the hill that functions as the basement of the building.

来自全球各地的二十名建筑师受邀为中国南京佛手湖畔各自设计一幢五房的度假小屋。我们的提案仿若传统染坊的结构,外部悬挂着布质帷幔,随风摇曳,掩隐了建筑物的框架,创造出模糊不定的形态。

靠近佛手湖轩时,初见还无法一窥全貌;踏上蜿蜒廊径,穿过秋千、锈蚀兽笼和焚松的火炉后,透过林间缝隙,湖轩才逐渐显露。在廊径尽头湖轩旁,终可一览佛手湖风光。在地面入口处,可见到五个方形木盒,如树屋般悬挂在空中;盒子两侧悬挂栗色布幔。从外部看,依稀可见各个木盒的独立楼梯。长形的水池倒映着建筑物的影子,视觉上与湖心连接。至于厨房、餐厅和客厅等公共空间,则半埋于地下室的山坡内,从外面不易窥见。

HAN-GU

函谷山庄

Beijing, China | Completion 2015
中国 北京 | 2015 年完工

VILLA

LOCATION	Beijing, China
CLIENT	China CYTS Tours Holding Co., Ltd.
FLOOR LEVELS	3 Floors, 1 Basement
BUILDING STRUCTURE	Steel Structure
MATERIALS	Composite Wood, Aluminum Panel, Stone
BUILDING USE	Villa
SITE AREA	31373 ㎡
TOTAL FLOOR AREA	10236 ㎡
DESIGN INITIATIVE	2011
COMPLETION	2015

项目位址	中国 北京
业主	北京古北水镇旅遊有限公司
楼层	地上 3 层、地下 1 层
建筑结构	钢骨结构
材料	环保塑木板、铝板、石材
用途	旅馆
基地面积	31373 ㎡
总楼地板面积	10236 ㎡
设计起始时间	2011
完工时间	2015

Han-Gu Villa is a 72-room high-end resort hotel. This project is located in a beautiful 600-meter long valley running east-west. From the site, one can see one of the most precipitous sections of the Simatai Great Wall. The valley is surrounded by abundant original vegetation, with a small stream running through in the middle. In order not to destroy its natural ecology and terrain, as well as to optimize the views to the mountains afar, the architect decided, with the client's consent, to keep the original landform and not to conduct any land formation works; the locations of the buildings will have to avoid touching the preserved tree and rocks, "tip-toeing" through the valley, as the architect puts it.

As the result, the buildings are scattered throughout the valley on stilts. The 8 clusters of guestroom units are carefully placed on the north and south side of the stream. Instead of common horizontal corridor circulation for the guestrooms, each of the 8 clusters has its own vertical circulation, and an open zigzag scenic bridge connects all of them following the creek. As they walk along on the elevated bridge, visitors are given optimal views of the Great Wall and the mountain landscape around the site. The scenic bridges overlap and crisscross the accommodation units and the creek, forming an interwoven dialogue between the man-made structures and the natural environment. The reception hall/club house is located at the west end of the site, acting as services and management facilities between the outside world and this secluded villa. On the exterior, Han-Gu Villa uses environmentally friendly composite wood with mixed natural wood color tones, echoing the constant change of sunlight and nature.

函谷山庄是一座 72 个套间的高品质休闲酒店，基地坐落于一由西向东约 600m 长的优美狭长的山谷中。在其中，可以相当清晰地远眺长城最险峻的司马台段。山谷景色优美、植被丰富，中间有小溪潺潺流过。为了不破坏其自然生态与面貌，并达到最佳的山景视野，建筑师的建议获得业主的首肯：完全不进行整地、尽可能保留原始植被与一草一木，建筑配置避开大树、大石，有如"踮着脚尖走过"此山谷一般。

因此，建筑组群顺势分散配置于基地中，采用基础架高形式建造。客房由西向东分为 8 个簇群规划，建筑韵律地错落在山谷内南北两侧。建筑的动线规划跳脱一般水平贯穿连廊的形式，而以一座架高开放的九曲连桥沿溪串联 8 组垂直动线。访客在桥上步移景异，又与客房单元及小溪相互迭穿越，在移动之间给旅人体验长城山景最好的角度，也形成人造建筑物与自然环境交织的精彩空间对话。接待会馆机能独立，配置于基地西端，作为内外动线的服务与控管。外观设计上，函谷山庄采用同中有变的原木色系，着重木材的纹理再现与细致的环保塑木板，多色掺杂搭配以呼应阳光与自然的多变。

Site model, aerial view
基地模型，鸟瞰

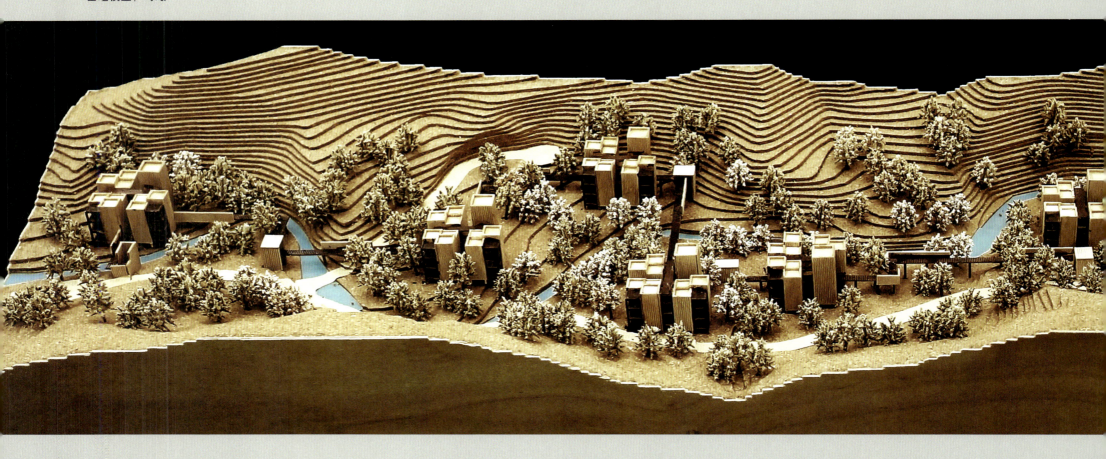

Concept sketch
概念草图

Guestroom unit elevation

客房单元立面图

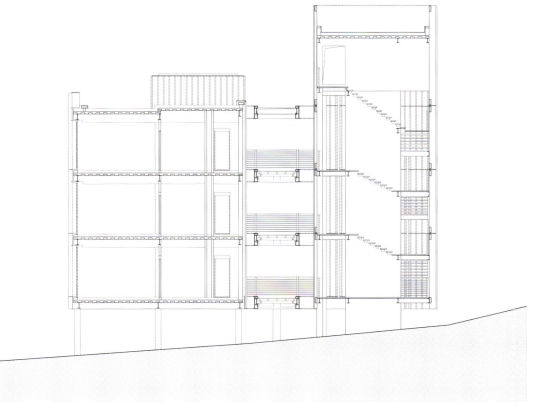

Guestroom unit section
客房单元剖面图

View from the villa to the Simatai Great Wall
由基地远眺司马台长城

Connecting footbridges
步道桥

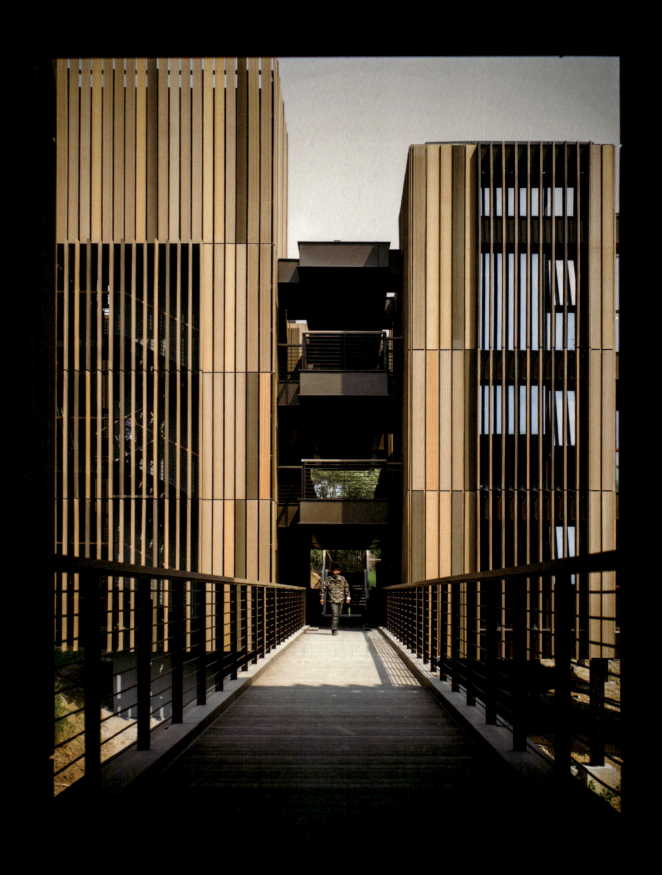

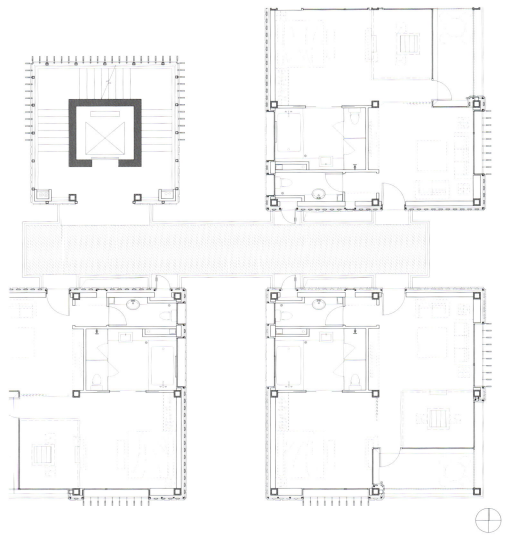

Guestroom unit plan
客房单元平面图

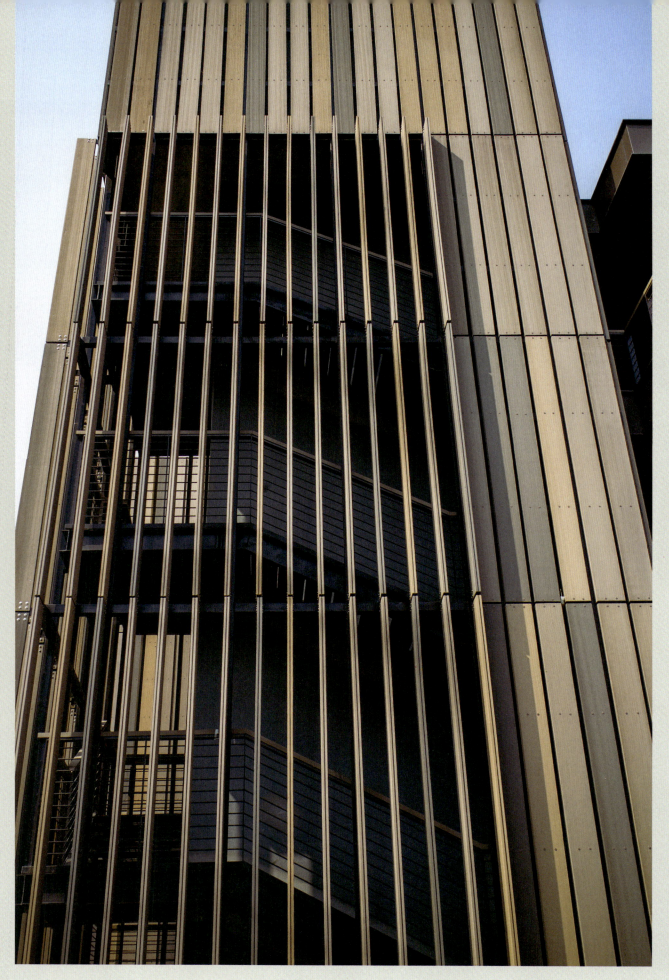

Facade details
立面细部

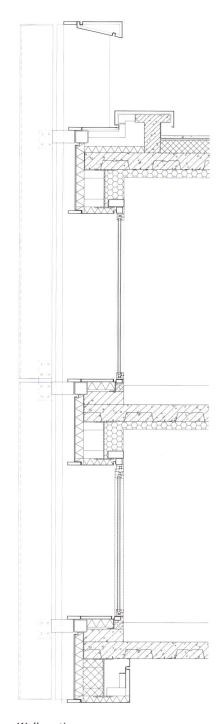

Wall section
墙身大样

NTU COSMOLOGY CENTER

台大宇宙学馆

Taipei, Taiwan, China | Completion 2017

中国 台湾 台北 | 预计 2017 年完工

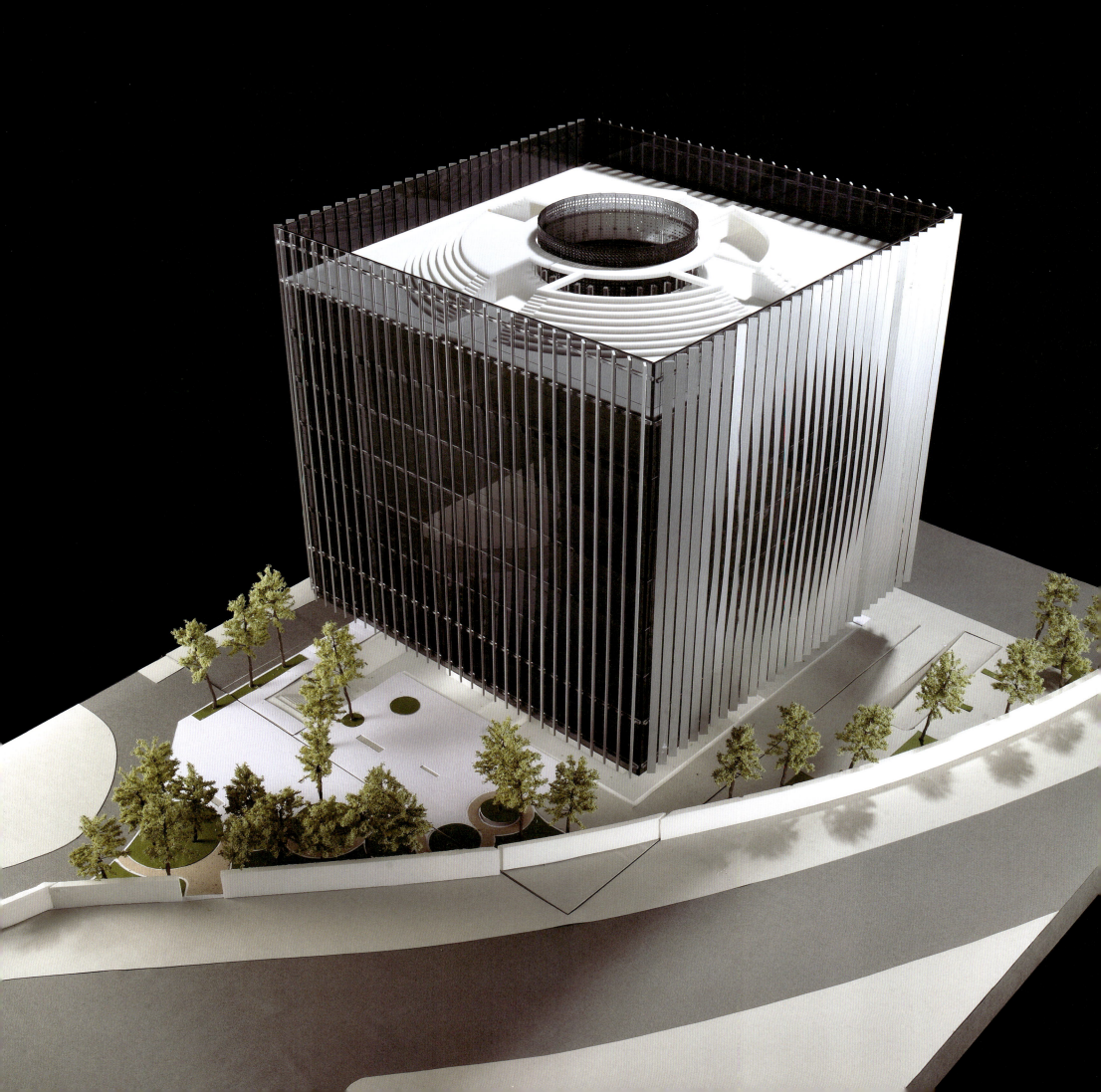

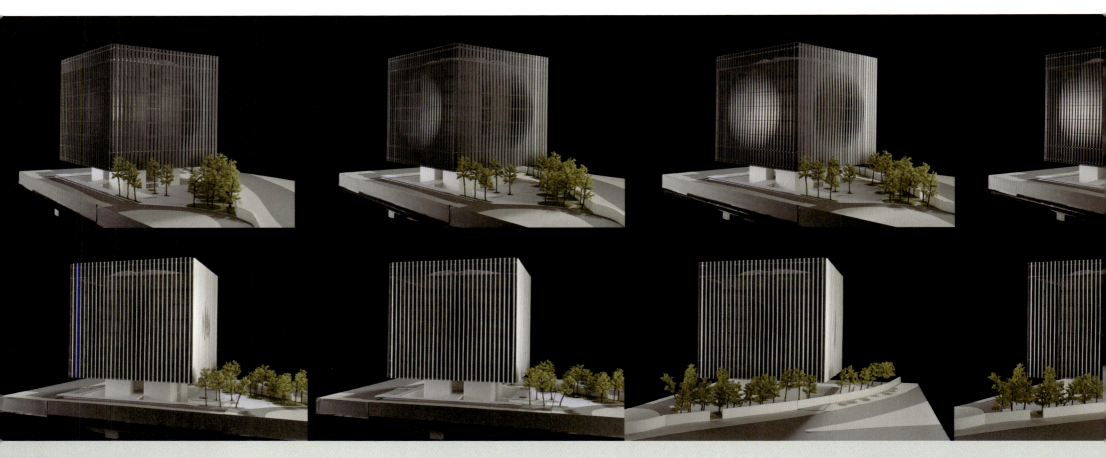

Model, showing different views toward the building
模型，从不同视觉角度观看建筑外观

LOCATION	Taipei, Taiwan, China
CLIENT	Chun-Yee Cultural Foundation
FLOOR LEVELS	8 Floors, 1 Basement
BUILDING STRUCTURE	Steel Structure, Reinforced Concrete
MATERIALS	Concrete Finish, Aluminum Panel, Glass
BUILDING USE	School Facility
SITE AREA	3115 ㎡
TOTAL FLOOR AREA	10913 ㎡
DESIGN INITIATIVE	2012
COMPLETION	2017

项目位址	中国 台湾 台北
业主	财团法人震怡文教基金会
楼层	地上 8 层、地下 1 层
建筑结构	钢骨结构、钢筋混凝土
材料	混凝土完成面、铝板、玻璃
用途	学校
基地面积	3115 ㎡
总楼地板面积	10913 ㎡
设计起始时间	2012
预计完工时间	2017

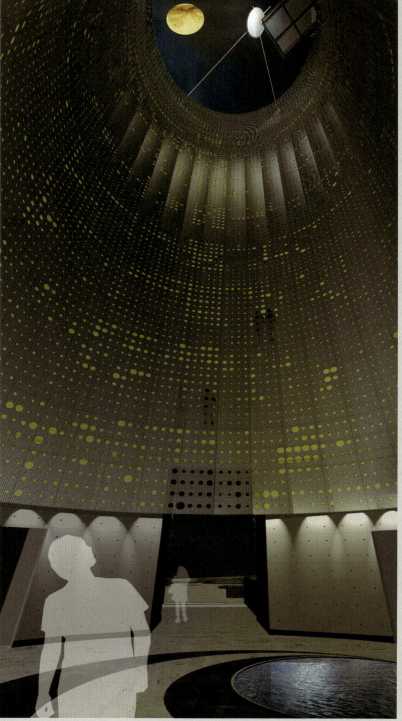

Courtyard day and night
中庭日景及夜景图

Seeking harmony with the surrounding environment, the entrances of the Cosmology Center extend outward from a cross-shaped axis. The plaza on the eastern side reaches into the existing banyan tree park. The idea of acting against the force of gravity inspired the design of a floating cube in space, structurally supported by the concrete core setback in the center. The depths of the vertical sunshades vary in a progressive sequence, so that the illusion of a sphere inside the cube can be seen outside as one moves around the building, visually experiencing a dynamic, changing facade.

Behind the external hidden sphere is a 38-meter high tubular open atrium, its height echoes with that of the Pantheon of Rome. It provides a direct communication between inside and outside, so that users indoors can directly sense the external natural environment, rain or shine, day or night. The interior facade of the atrium models the heavenly bodies, made by perforated metal claddings. It also provides visual penetration through the hallways and decreasing echoes in the atrium.

The second through eighth floors are for laboratories. There's an outdoor viewing terrace on the seventh floor for relaxation. The Cosmology Center's simple design also cleverly includes a number of green building features.

台湾大学宇宙学馆配合周边环境的和谐，建筑物出入口以十字轴向外延伸，于东侧设置广场，接续现有的榕树广场；借由反重力托起的概念，设计以内缩之剪力墙将正方量体撑起，创造出视觉上悬浮的立方体；并以"隐含的球体"作为立面设计概念，由垂直向度不同宽窄序列渐进的遮阳板创造出逐渐现形的天圆，当人们移动观看时，即可感受到立面动态、渐变的视觉经验。

外在隐涵的方圆之中，挑高 **38m** 的观月中庭天井，一如罗马万神殿的高度，连通室内外自然的流动，让内部使用者能直接感受到外部自然环境的晴雨·昼夜。中庭内立面以寰宇星体的罗列为概念，透过金属冲孔板的设计呈现；同时适度提供通廊的视觉穿透性、降低空间压迫感及减少中庭回音的现象，也增加各楼层间使用者的互动机会。

二层至八层作为研究室使用，于七层设置户外观景露台，提供给日常学术研究人员一处身心休憩平台。宇宙学馆简洁的设计巧思亦包含了许多绿建筑理念。

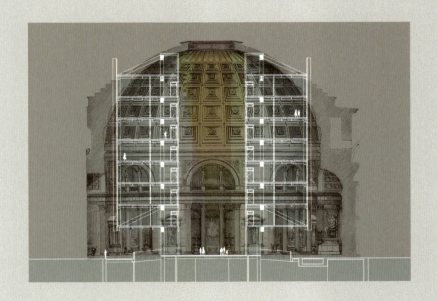

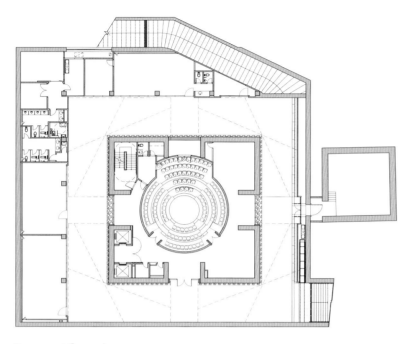

Basement floor plan
地下层平面图

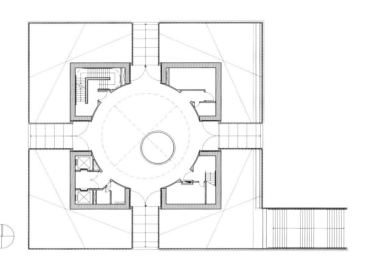

Ground floor plan
地面层平面图

South elevation
南向立面图

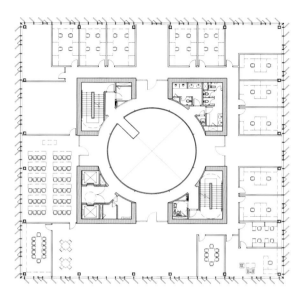

Typical floor plan
标准层平面图

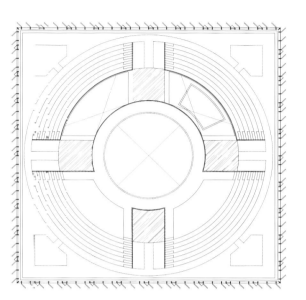

Roof floor plan
屋顶层平面图

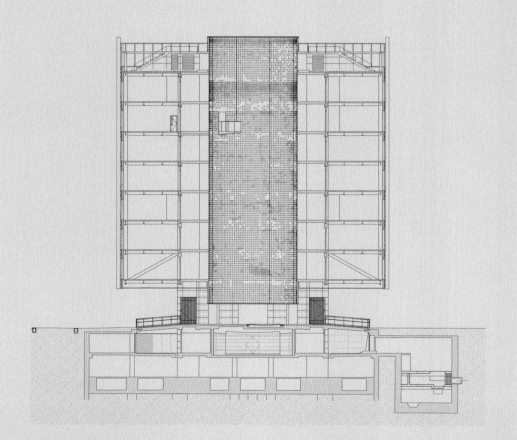

North-south section
南北向剖面图

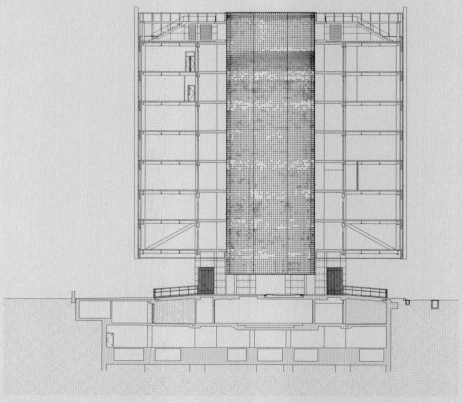

East-west section
东西向剖面图

WUTAISHAN RETREAT CENTER

五台山禅修中心

Shanxi, China | Completion 2017

中国 山西 | 预计 2017 年完工

LOCATION	Shanxi, China
CLIENT	Wutaishan Zhenrong Yuan
FLOOR LEVELS	5 Floors
BUILDING STRUCTURE	Reinforced Concrete
MATERIALS	Grey Brick, Textured Concrete Finish, Glazed Curtain Wall, Steel Plate
BUILDING USE	Retreat Center
SITE AREA	16650 ㎡
TOTAL FLOOR AREA	6685 ㎡
DESIGN INITIATIVE	2013
COMPLETION	2017

项目位址	中国 山西
业主	五台山 真容寺
楼层	地上 5 层
建筑结构	钢筋混凝土
材料	灰砖、粗纹混凝土墙面、玻璃帷幕墙、钢板
用途	禅修中心
基地面积	16650 ㎡
总楼地板面积	6685 ㎡
设计起始时间	2013
预计完工时间	2017

North-south section
南北向剖面图

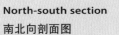

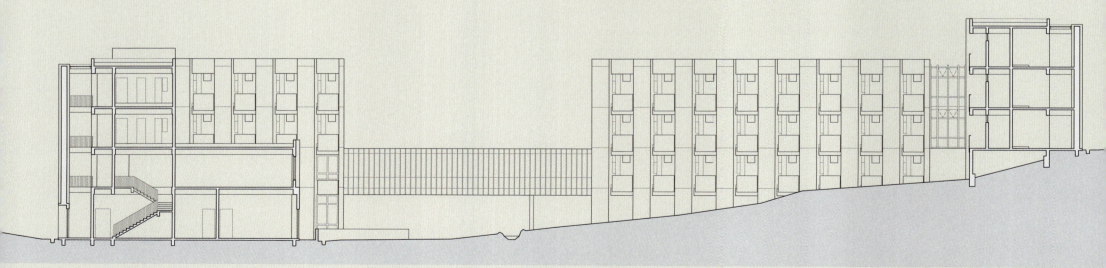

East-west section
东西向剖面图

The retreat center is located in a valley of the well-known Buddhist holy site: the Mount Wutai. It has a gorgeous and serene setting, surrounded by mountains with a stream flowing by on the eastern side and a line of apricot trees running across the sloping terrain.

The design is a simple square courtyard building being place on the sloped site. As it sits on the slope, the perfect square starts to shatter; openings from northwest and southeast are placed to allow the apricot trees to "pass"; the remaining parts of the square also settle on different levels of the site so they are broken and shifted. The result resembles the image of a broken square plate on the ground.

All the retreat rooms look into the serene courtyard, with corridors and stairways on the outer peripherals of the building. These sky lit communal spaces have occasional breaks at where the square has been broken or shifted, allowing glimpses to the views outside. The massive thick walls surrounding the building are veneered with local grey brick, with small openings to bring in the change of sunlight. On the courtyard side, rough architecture concrete is used, with punctuations of wooden individual balconies for each room.

此禅修中心坐落于佛教圣地——五台山区——的山谷内，三面环山，东侧有溪水流过，环境优美寂静，还有一行古老的野山杏自基地坡面自上而下穿过。

建筑设计是从一座单纯的方形四合院放置在基地的山坡地上开始的。当此合院放置于坡地上时，完整的方体开始剥裂：西北与东南两个角隅开放出来，以便野山杏"通过"；方体落在不同高低的基地上，也因此断裂或错位。结果是一个状似破碎在地上的方盘子一般。

所有的禅修房间都面对内庭，走廊与阶梯则设置于建筑物的外围。这些采天光的公共空间在建筑物断裂或错位处都设有休憩角落，人们在此可以一瞥室外的景色。厚实的外墙以当地的灰砖包复，墙上有小开口，引进日间光影的变化。内庭立面施以木纹粗面清水混凝土，点缀着各禅修室装有木栏杆的阳台。

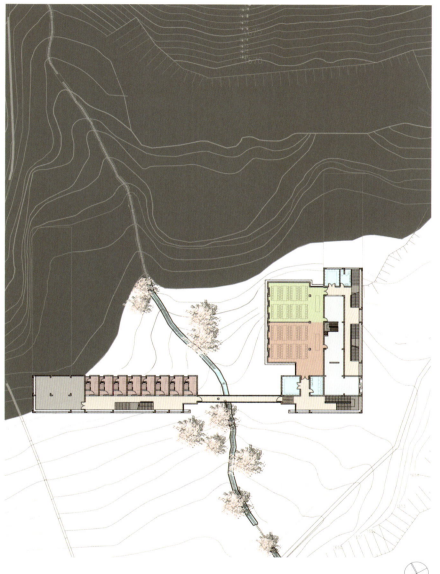

EL+4.200 Plan
EL+4.200 平面图

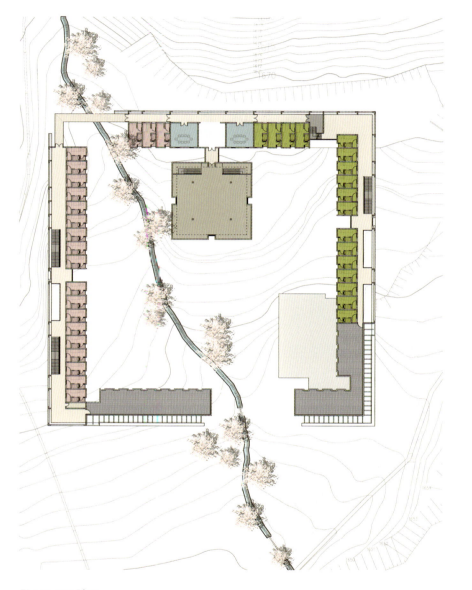

EL+14.100 Plan
EL+14.100 平面图

Site plan
基地配置图

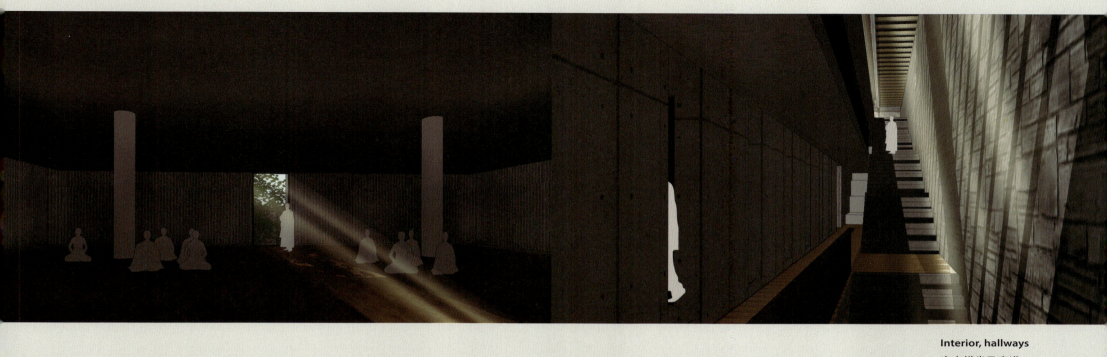

Interior, hallways
室内禅堂及廊道

ACKNOWLEDGEMENT
感谢

Established in 1985, KRIS YAO | ARTECH marked its 30-year milestone in May 2015. On this special occasion, we selected 30 projects over 30 years, inspiring the exhibition name "30x30". We are very grateful for the Urban Environment Design Press and Liaoning Science and Technology Publishing House for jointly publishing this 3-volume monograph.

Architecture involves a myriad of complicated layers, and cannot be completed by just a few select individuals. Rather, it requires numerous experts from various specialties, a tremendous investment of effort and time, and a continuous cycle of iterations to complete a work. The lifecycle of an architectural project is complex: from thorough communications with clients and users in the initial stage of design, to detailed cooperation with consultants during the design process, to meticulous coordination with builders in the construction phase. Therefore, while 30 years may sound like a long time to be an architect, it is not incredibly long when we look through the perspective of architectural projects. In the current fast-paced world, architecture is one of the rare and precious specialties that require maturation through time and patience.

I would like to extend a special thank-you to our clients from whom our designs are inspired by. Their continuous support and generosity has enabled KRIS YAO | ARTECH to grow and mature over time.

Architects are like film directors; they need to collect, synchronize, and harmonize different talents to produce a great architectural cinema. Therefore, I would also like to express my gratitude to all of our consulting teams across all fields (e.g., structure, MEP, landscape, civil, lighting, acoustic, exterior wall) for their significant contributions. My gratitude goes to many friends in academia and to fellow architects, for their continuous suggestions and critiques. Lastly but certainly not the least, to all the co-workers throughout the last 30 years, for their unwavering perseverance and pursuit of shared values, which have made many projects possible.

I want to thank my wife, Xiang, for her continuous inspiration for my learning and improvement, and her unwavering and unconditional love, tolerance and support; to my three beloved children, Joyce, Julian, and Adrian, watching all of you growing up, is my upmost source of happiness and gratitude. Finally, my sincere respect goes to my spiritual master, Dzongsar Jamyang Khyentse Rinpoche, for being a constant and compassionate lighthouse throughout my life journey.

Kris Yao, Hon. FAIA
2015 / 6 / 16

姚仁喜 | 大元建筑工场成立于 1985 年，今年适逢 30 周年的执业里程。我们回顾并挑选了 30 年来的 30 件代表作品，举办题名为 "30X30" 的展览。辽宁科学技术出版社，为此出版了这套名为《姚仁喜 | 大元建筑作品 30x30》的精美作品集，分为《艺》、《聚》、《思》三册。在此特别向出版社诚挚致谢。

建筑牵涉的层面既广泛又复杂，无法靠少数人就能完成，需要集合众多各种领域的专业人才，投注大量的心力与时间，反复淬炼才能有所成果。建筑作品从设计初始与业主或使用单位的密切沟通与互动，到设计过程与顾问团队的切磋咨询，到建造期间与营建单位巨细靡遗的协调整合，无一不是繁复耗时的工作。对建筑师而言，30 年算是很长的执业岁月，但相对地从作品完成的观点而言，30 年却不是太长的过程。建筑设计是少数需要时间与耐心才可能完成的行业，在当今一切讲究极速效率的世界中，它是值得珍惜的。

我要感谢 30 年来启发我们设计理念的业主，由于他们不断的提携与支持，使我们得以茁壮成长。

我要感谢 30 年来国内外的专业合作团队，包括结构、机电、景观、大地、灯光、音效、外墙等各种领域，所给予我们的协助。感谢学界及同行友人，给予我们不断的建议与指教。更要衷心地向 30 年来一起工作过的同仁们致敬，经由我们共同的努力与坚持、团结力量的凝聚与共同价值的追求，才能让许多作品有令人满意的呈现。我常比喻，建筑师的工作有如电影的导演，需要汇集并整合各种优秀的大家，才能演绎出精彩动人的建筑剧码。

我要感谢我的内人任祥，30 年来持续地激励我，促使我学习与成长，而且无条件地给我容忍与支持；感谢三个亲爱的孩子：姚姚、JJ 与小元，看着你们成长，一直是我感恩与快乐的源泉。最后，我要献上最诚挚的敬意给我的精神导师——宗萨蒋扬钦哲仁波切，感谢他作为我生命旅程中永远慈悲的明灯。

姚仁喜 建筑师
2015 / 6 / 16

图书在版编目（CIP）数据

Social Sanctuaries 思／姚仁喜著 ． — 沈阳：辽
宁科学技术出版社，2015.8
ISBN 978-7-5381-9371-8

Ⅰ．① S… Ⅱ．① 姚… Ⅲ．① 建筑设计一作品集一中
国一现代 Ⅳ．① TU206

中国版本图书馆 CIP 数据核字 (2015) 第 176953 号

姚仁喜｜大元建筑作品 30x30

著　　者 :姚仁喜
编辑总监 : 刘玉贞　　姚任祥
编辑执行 : 温淑宜　　乔　苹　　林宜熹
美术总监 : 段世瑜
美术编辑 : 方雅铃　　陈怡茜　　郑乃文
摄　　影 : 刘俊杰　　郑锦铭　　游宏祥　　潘瑞琮　　马怀仁　　陈弘昕　　张基义
　　　　　刘振祥　　邓博仁　　李东阳
策　　划 : 彭礼孝　　柳　青

出版发行 : 辽宁科学技术出版社
　　　　　（地址 : 沈阳市和平区十一纬路29号 邮编 : 110003）
印　刷　者 : 北京雅昌艺术印刷有限公司
幅面尺寸 : 300mm x 300mm
印　　张 : 16.5
字　　数 : 20千字
出版时间 : 2015年8月第1版
印刷时间 : 2015年8月第1次印刷
责任编辑 : 包伸明　　张翔宇
责任校对 : 王玉宝

书　　号 : ISBN 978-7-5381-9371-8
定　　价 : 160.00元